Dear Taylor
love from your Grandpa

[illegible signature]

An impression of the book
"An American in the Mountains of Iran"

Close friends, Masoud Akhbari from Iran and the Farsi speaking American Blaine Taylor, wrote a book that is a must read for everybody looking for the genuine Persian soul over the centuries.

The beauty of their common writing lies in the detailed descriptions of hunting activities and meals together with the Bakhtyari , genuine mountain and desert nomads, at the end of the twentieth century. It brings one back in another world not found in the cities. We have no idea how these people presently cope with hardship in an unforgiving nature. Incredible that the two authors experienced this so recently.

The book gives also a good picture of meetings with customs, police and army officials in presentday Iran. Might change possible prejudices.

Jan J.M. Muldrink
General Manager for several business units AKZO NOBEL
Managing Director AKZO Research Lboratories Arnhem.
Chairman of Royal Dutch Chemical Association Board 1994-1996
Chairman of DCO(Sustainable Chemistry Developement) 1995-2005
(DCO an association of industry and government to develope sustainability in enegy and raw material).
Director in enterprises for manufacturing process equipment

An American in the Mountains of Iran

An incredible, touching story - straight from the heart - of adventure, friendship and brotherhood across cultures.

It makes you want to travel the beautiful Zagros Mountains of Iran, too, and meet the proud and independent people living there. Preferably with a guide like Mahmeed.

In this well-written tale the authors, Masoud Akhbari and Blaine Taylor, shed light on a bit of recent history and give us hope for a more peaceful tomorrow, based on respect, friendship and understanding.

Freek Eland
Editor-in-chief Radio Netherlands Worldwide, 2000-2004
Senior lecturer Intercultural Communication, Rotterdam University of Applied Sciences.

Masoud and Willie

Blaine, Judith and Family

An American
in
the Mountains of Iran

The story of two people from different cultures and religions, who became brothers in a relationship which has lasted over four decades.

By: Masoud Akhbari
and
Blaine W. Taylor

First edition 2013

Printed in China

ISBN 978-0-9890829-0-7
Library of Congress Control Number 2001012345

Published by: Taylor Family Enterprises, LLC.,Cheyenne, Wyoming, USA

PREFACE

Preface written by Dr. John Godrich; Minehead, Somerset, England.

An American in the Mountains of Iran

This account of Big Game hunting in the Zagros mountains of Iran is unique, most exciting and very informative.

It is written by Blaine Taylor, a geophysical engineer from in the USA and Masoud Akhbari, an Iranian engineer who used to work for Dutch Shell Petroleum Company at Ahwaz in Mesopotamia before the Nationalization of the oil wells.

Together with two local guides they set out on horseback in 1974 to explore the high mountains of the Zagros. These rise to a maximum height of 4500 Meters (nearly 15,000 ft.) and run from a point 200 miles North of the Persian Gulf for 500 miles to the North-West parallel with the Iraqi border.

Masoud, who now lives in Holland with his Dutch wife, helped Blaine to get permission from the government to import his superb rifle through the Customs of a Country where firearms are strictly forbidden in private hands.

Blaine had been hunting all his life in the United States and had become a crack shot, with boundless knowledge of tracking wild animals and the complicated skills of repair and maintenance of rifles and shotguns.

In the Zagros highlands they hunt for long-horned mountain goats (Ibex), gazelles, wild boar, brown and black bears, wolves and many sorts of game birds. In the summer, the Bakhtyari Nomads live there in their woolen tents herding their sheep and goats.

The tale tells of a life high up in the mountains, riding two good mules, living rough, and sometimes enjoying food and drink with the local people in return for providing them with meat or removing a threatening wolf.

Closely dependent on one another, often in danger, crossing a raging river, nearly getting killed by a bear and falling off a precipice, even the hostility of suspicious villagers, they share this perilous expedition and become sure, trusting, lifelong friends. Our comparatively secure, safe, organized lives are in marked contrast to the wild world these hunters shared together.

Acknowledgements:

For the patience of angels, our wives, Willie Akhbari, and Judith Taylor, who not only put up with our traveling forays, but also encouraged us to write this book.

To: Dr. Hormuz (Harry) Fathi, who accompanied us on many of the trips, and supplied some photographs to us.

To: Dr. John Godrich M.A., M.B. B. Chir (Cantab), Author of his father's diary of 1908-1919 in the WW1 called "Mountains of Moab"

To: Mr. Simon Tuinder, who has been in Iran working with Helicopter Services for providing some fine photographs used in our research and writings.

To: Mr. Masoud Nemati, an Iranian geologist, who verified some later backup pictures of Dasht-e-gol, and Kuh-e-Safid, after 40 years of elapsed period from our initial travels.

To: Mr. Human Almassi, who provided us with photographs of Bakhtyari people in recent times.

To: Mr. Vali Moshk Alizadeh, Tehran, Iran who provided us with the pictures of Bakhtyari Khans and their 1907 involvements of Iran's constitutional revolution.

The Yearly Migration

The Bakhtyari nomad tribes of Iran are a simple but hard and tough people. In general the Bakhtyaris live six months in their “Qeshlaq” (winter quarter) around Masjid-I-Sulaiman on the Khuzestan plain and four months in their “Yaylaq” (summer quarters) spreading on the Zagros Mountains and close to Shahr Kurd, and Esfahan, which the Bakhtyaris call it “Charmahal va Bakhtyari”. The other two months of the year they are migrating and grazing their flocks on the way. The Bakhtyaris are a very independent people, like most other tribal unions, they are composed of various clans with common family links. Each branch of the Bakhtyari tribes is ruled by strong ruler called “khan”. While fairly widely dispersed in the summer, which prevents much social intercourse, the winter spent on the Khuzestan plain provide ample opportunity for the extra-familial socialization and arranged marriage which prevents the great scourges of genetic problems which come from prolonged periods of inter-marriage.

The contrast between the modern world and the way of the life the nomads of Persia, who live and work hard to survive in those great mountains without modern conveniences, transportation and medicine, is truly fascinating. To cross the rivers they throw themselves on inflated goat skins and paddle until they reach the opposite shore. Some of them cross the river 6 to 8 times a day to and fro to help the elderly carry their flocks, children and household utensils to the other side of the river. Crossing each river may take a week resulting in much causality to their livestock, which are swept away in rapids and whirlpools. Even more sadly the cost is sometimes higher; those who drown trying to save their friends and neighbors, and preserve their way of life.

During these annual migrations, the Bakhtyari cross several high mountain ranges. Not only are these mountains a great hardship, but the verdant valleys provide the sustenance for their flocks to enable them to make this month long transition.
As with most migrations, the path between summer and winter grazing would not support a long occupation. It is being entirely exhausted by the flocks in the short time of the migration.
“Zard Kuh” (literally yellow mountains), in the central Zagros range is about 14000 ft. high and is still covered with ice and snow when they start their seasonal migration.

Friends for this life and the next
Blaine W. Taylor and Masoud Akhbari

Contents

An American in the Mountains of Iran

Chapter 1

Life Before Going Abroad

I was born in a small town in southern Idaho, by the name of Malad. That name comes from the French Malade, meaning "sick". It seems that the early French explorers when camping in the valley, had become very sick, some even dying when they camped there for a couple of days.

Malad valley is a beautiful, green, mountain valley, with grass that grows as high as a horse in the summer, and a beautiful stream running through the middle. It eventually dumps into the Great Salt Lake, which is the modern remnant of an inland sea which was called Lake Bonneville in ancient times. Lake Bonneville broke out and most of it ran into the Pacific Ocean about 15,000 years ago, just before the last ice age finally receded. This period of glaciation left many of the U shaped valleys we see in North America today, and shaped the face of what we now call home.

Part of the glaciation carved out mineral deposits from the rocks. One of those mineral deposits was beds of arsenic. The Malad River runs across several beds of arsenic, and after dissolving that poison, becomes extremely toxic to those who drink the water.

I feel a real distinction in being born and raised in "sick city Idaho".

This whole area for thousands of square miles is subject to poisonous vapors and mineral deposits. The majority of Yellowstone National Park's Firehole River is not good to drink because of all the geysers spewing toxic mineral water into the area.

My father, and his father before him were blacksmiths. Grandpa's specialty was shoeing horses. My father was a specialist in working metal, and making mechanical repairs to farm machinery. When young, I actually worked as an apprentice to my dad, and learned the trade. Dad didn't like shoeing horses, and so we didn't do much of that. What little we did, I had to do, because the horses sensed that my father didn't like them, and they would bite and kick instead of calmly standing by when being shod. You would probably not think about a horse biting, but they bite like an alligator. I have seen them crush a full grown wolf's skull by biting them on the head. When fighting wolves in the wild, this is a horses' preferred method. They certainly don't want to rare up and open their soft, un-protected underbelly to a wolf.

My father and I were quite close, although because of my sickly conditions as a young child, I always felt that I couldn't come up to his expectations. My health suddenly became good when I turned 8 years old.

Dad had the best vision of any man I have ever known. Before his death, he could read a newspaper across the room. The only problem he had as he got older, was that he couldn't read it up close. He was always complaining that his arms were getting too short.

Dad and I did a lot of outdoors stuff together. He liked to fish more than anything, and I liked to shoot. When I was about six, dad would leave work in the evening, and we would drive together to a small creek where he would fly fish. He would always catch 8 or 10 small trout. While he was fishing I was using a bow and arrow to shoot flicker-tail gophers. We called them "ground squirrels". The farmers loved for me to come and shoot gophers, because their horses and cattle were often stepping in the holes, and breaking their legs.

When I got older, I substituted a .22 caliber rifle for the bow and arrow. I never did take to fly fishing like my father. One of my treasured memories was stopping at the gasoline station on the north side of town, and having a "Barq's" Rootbeer or Orange soda, while we were waiting for someone to come in, so that dad could give them some fresh caught trout. If we had eaten every fish my father caught, we would have had gills!

Malad is a funny little town. The area was settled almost entirely by Welsh immigrants. They were extremely hard working people, but very clannish. My father and mother had moved in from a different part of Idaho during the second world war, when the town needed my father's services as a blacksmith, and we didn't have any family there. We just didn't fit in, and so I was raised to be a loner. Those traits stick to me to this day.

My father was born in 1911, and the great depression began in 1929, with the stock market crash. My father had only just become 18 years old when this event happened. His entire adult life was shaded by this terrible depression, and he never got over it. Mom and dad never wasted anything. As soon as the great depression was over, we went into the rationing of the second world war. Because dad was a blacksmith and so integral to that farming community, the community wouldn't let him go to war. He always felt diminished when all the other men went to war, but he stayed home.

I remember mother and dad collecting cigarette package paper wrappers, and separating the aluminum foil from the paper. They wrapped these pieces of foil into a big ball, and every so often they gave them to a group for collection and making this back into airplanes. What a mark those hard times put on everyone that lived them.

My grandfather gave me my first .22 caliber rifle. I treasured that gun, and recently passed it on to my youngest daughter for her children's posterity. I believe I was 8 years old when grandpa gave me that gun.

That gun is a Remington pump repeater. I had to go out with my father, and practice with my air rifle, until I could pass his exam and be granted the ability to take my .22 rifle out alone. This .22 rifle would hold 15 .22 caliber long rifle cartridges, and the test was that at 50 feet, I had to light 15 parlor matches, without reloading the gun. When I was 10, I passed the test. I could also shoot small plastic prescription bottles out of the air when thrown.

Dad was unforgiving about missing a shot on a deer, or wounding an animal. I was taught to be sure of my shot so that the animal didn't suffer, and to ensure I was ready to take the shot before trying to do so. Because we were poor, virtually all of our meat was wild meat, fowl, and fish. We always took as many deer as the hunting seasons and regulations allowed us, and fishing in the summer put a lot of fish away for us.

There were not a lot of people who had home freezers. Only the rich could afford such a luxury. We always had a rental "Locker" for frozen food at the locker plant. A locker plant was a cold storage room in a meat processing plant. They had the big cooling equipment, and they made an area available to the people of the town with big drawers to store frozen food. They rented these drawers, which were about the size of a large home freezer. We had a lock on our drawer, and every time we needed frozen food, we went to the locker, which was only a short distance from our home. There we asked for our key, and took what we needed for that day, or week, and left. This same processor skinned, and cut up our wild game for us. They would add commercial port to the very lean meat to make sausages, and their costs were very reasonable. They skinned the animals and didn't charge us if they kept the skin.

We had a very large garden, and took fresh vegetables out of it all summer long. My mother spent long hours bottling and canning vegetables, which lined the shelves of our cellar. Those vegetables lasted us through the winter and spring, until we were ready to start harvesting the next year's crop. Although we were quite poor, I think we ate better than the rich in the big cities.

Every year in the spring, our neighbor who was a farmer, would bring his tractor by, and in a matter of about an hour, he would plow up the entire garden plot of about an acre for us. I really liked that man, because he was just a good man. He lived right next door to us, and never married. Every evening in the summer and autumn, he would sit out on his front porch and have a cigarette before going to bed. I seldom missed a chance to go and sit with Mr. Lusk, and talk with him. Since we were both loners and kindred spirits, we didn't have to talk much, but just enjoy each other's company.

My father had developed a process of creating ornamental iron railings for concrete porches. Dad sold these around the area, and was quite famous for doing so. When I had sufficient skill, the first set of porch railings I built, were installed on Mr. Lusk's front porch. I'll never forget the tears in his eyes. He didn't go to church like most everyone else in that little town, he, like our family preferred to let his Christian actions say how he felt about his relationship with Deity.

When I was 13 years old, I decided to take a seminary class. I had been born into a Mormon family, but we were not active participants in weekly religious observance. I decided that I liked this particular man in the Church, and since he was the seminary school teacher, I would like to learn more. I took his class in the study of the Old Testament. During that year, I fell in love with the Old Testament. When we later moved to Israel, I had a tutor teach me the Hebrew language out of the Old Testament. I was often told what funny Hebrew I spoke because of the origins of my tutelage.

This teacher, Jack Raymond, changed my entire life with his gentle honorable ways. I started going to church, and gave up alcohol and tobacco for a time. He was also the 9th grade basketball coach. I participated on that team, and did well, as I did in all athletics.

I played high school basketball, for two years, but had a falling out with the basketball coach, and left the team during the first game of my last year in high school. When I quit playing basketball, the entire church team who were being coached by my old teacher and friend, Jack Raymond, approached me and asked me to play with the church team.

The only conditions were that I had to regularly attend church, just like the other boys, and give up smoking and alcohol.

The Mormon faith has a "Word of Wisdom", or health code accepted by the members as a commandment and commitment. This Word of Wisdom proscribes use of Alcohol and Tobacco, and requires the use of grains as the bread of life, with vegetables, fruits in their season, and the sparing use of meat. Killing animals is a sacred thing with Mormons, and their use as food is a requirement if an animal is to be killed. Indiscriminate killing for sport is looked down on, and I think reflects the values of the American Indian which was a western heritage. Only kill what you need, and make sure you need what you kill.

I remember when I was 16, I decided to take a trophy deer, instead of looking for a really good piece of meat. I hunted hard and long, finally taking a huge 6 point buck, on the last day of the season. In the west, we count only the points on one side, in the south this would have been counted as a 13 point buck. It had 6 on one side, and 7 on the other. It also field dressed over 350 pounds. It was old, hard to kill, and when we tried to eat the meat, it was hopeless. I remember my mother telling me that "you couldn't stick a fork in the gravy" of that deer. We ground it up, and mixed it with pork to make sausage. I got so tired of sausage that year. I remember putting that rack of antlers on the end of the garage, and when we moved away, two years later, I didn't even bother to take it with me. It would now be record setting antlers, but that was not our way.

Jack Raymond and I went separate ways later, and when he moved to another town, and was drowned in a tragic fishing accident, it was a terrible shock. When I later became fully active in the Mormon Church, I often thought how much this man did for me, and what he meant to me.

I was a good student, and I won the Bausch and Lomb science award for my school. I didn't have the highest grades of my school, because of one class which I had a hard time mastering, but my grades were high enough to give me a scholarship to go to university.

During the last four years of high school, I worked each night from 1:00 am till 8:00 am in the morning, at a flour mill. After my work shift, I went to school. At 3:30 pm, after school was out, I practiced basketball or ran on the track team. When I finished basketball or track, I went home and slept until it was time to go back to work. I never thought that there was anything wrong with this. My family had the motto that it was better to wear out than rust out, and that is the way we lived.

I went to university in 1961, and met my beloved Judith at a "hello" dance in 1962. I fell in love with her literally at first sight, and have never been able to understand what she saw in me. I was a rough kid without many saving graces, except knowing how to work. We were married in the spring of 1963, and in 2013 will celebrate 50 years of married life.

Judith's Father and Mother, Judith, Blaine and Blaine's Mother at the wedding

In 1968, we took out an educational loan with the Southern Idaho Farmers Credit Coop. Rather than begrudging this debt, we were grateful for it because it allowed me to finish my education in Physics, Math, and Chemistry at Utah State University in Logan. I had been going to school part time for many years, and with a wife and two children to support, it was dragging on. The Loans allowed me to return to school full time and finish. My last year was a bit hectic even with the help that the loan provided, I had a part time job in the Physics lab, and carried 24 quarter hours of upper division physics and math classes that last year in order to finish and not prolong the pain. A normal load would have been 15 quarter hours, and it was a strain and constant tension. Judith also worked those last two years, so we could make ends meet in a city, which had a reputation for abusing students as far as wages were concerned. We were both paid less than minimum wage, which they could get away with because it was a University, and not subject to the same laws and rules as the private community. When I finished my university work, I had declined to a mere 130 pounds, and for a man of six foot three inches that was pretty small. I have never seen that weight since!

Blaine with Oldest Girls at College Graduation

After graduation, we moved to North Dakota, to work for Schlumberger Well Services. We often referred to that as our first foreign assignment. It was very different than where we grew up in Idaho and then Utah, where we finished university.

In North Dakota, we found the weather extreme. The summer was short, humid and hot, and the winter long, dry and cold. On the 19th of January 1970, I had an automobile accident, and the local newspaper published a story saying it was 70 below zero Fahrenheit, with a 40 mph wind blowing besides.

I had been returning on a Friday evening from an oil drilling rig, after having run well logs. When an oil well is drilled, the information gained is not very good. There are cuttings from the drill bit which are brought up by the drilling fluid circulating from the top, down through the drill pipe, out through the drill bit, and back up the hole. Those cuttings have a trip of sometimes as much as an hour from the place where they are removed, until they get to the top of the hole. The “lagging” of samples is quite imprecise, because the smaller ones go right up with the drilling fluid, or “mud”, and the larger ones don’t lag so well. Depending on the speed of penetration of the drill bit, there is quite a lot of doubt about where those samples actually came from. The answer is to drop electronic and nuclear tools down a well on a very long and strong electrical cable, and when pulling them back up, measuring the characteristics of the borehole.

When the logging job was finished, I was returning to base in Williston, North Dakota, in my car, with another colleague in the passenger seat. The wind was blowing hard, and the snow was blowing with it, impairing vision, in what was there called a ground blizzard. I was driving very slowly down the road, when I was hit hard from behind. My car was knocked off the road, and when we came to a stop, I immediately checked on my passenger who was unconscious. He was quite badly injured, and I suspected his neck was broken in the impact. I later found out that it was a good thing that I didn't move him because that was actually the case.

I then got out of the car, with some difficulty, because the doors were jammed. The entire back section of the car was pushed up into the rear passenger compartment. I went to see the occupants of the car which had hit me. There were beer bottles everywhere. There had been six occupants of the car. One who was sitting in the center of the front seat, was ejected through the windshield, and was unconscious on the hood of the car. Four of the other five occupants were also unconscious. The driver was the only one who was still conscious, and he was pinned under his steering wheel. I was the only one in either car who escaped injury.

We were about 2 miles from the nearest house, on a dirt road, and since it was the drilling crew just leaving the rig who hit my car, there was little chance that anyone would be along to help. I covered everyone up with my sleeping bags, and started out for the nearest house. It was so cold and the snow was blowing so hard that I just wanted to lay down and sleep in a snowdrift. When I realized what was happening to me, I quickened my pace. I remembered the "scout pace", which is to run or jog 50 steps, then walk 50 steps. A person in reasonable condition can keep this up on level ground all day. I began to make some real progress. At this rate you can cover a mile in about 10 to12 minutes.

I arrived at the first house, only to find it closed down, boarded up for the winter, and all the utilities turned off. I had decided that I would break into a house if I had to, since it was a matter of life and death, but there was no reason to waste time here, the occupants had moved from their farm into town for the winter, and they would have shut off phone, and all utilities, so I carried on.

I arrived after about 45 minutes at another house, which showed signs of habitation. I went up and knocked on the door. Minutes later a little boy of about 5 answered the door to see what I wanted. He told me he was all alone, and the house was cold. I quickly went in, used the phone to call the sheriff's office, and had them call children's welfare services. I also called my business, because I wanted them to tell my family I was not injured, and because I needed help in securing the radioactive source I had in the car. I gave them the directions to reach me, then I went about finding what was wrong with the heating system.

I quickly found that the heat was off, because the propane furnace's pilot light had blown out with the wind storm we were having. I got the pilot re-lighted, and about that time the sheriff, two ambulances and children's welfare people arrived. The ambulances were Volvo, and had shutters over the radiators for cold weather. The sheriff's car was an Oldsmobile, and it was blowing steam, because the radiator had frozen. We quickly packed snow in front of the radiator, to keep it from drawing air, and followed the ambulances to the accident scene.

To my utter amazement, everyone survived this tragedy. My colleague wore a neck brace for several years, and I believe he left the oil business because of the accident.

My third daughter was born two days later. I think the strain on my wife caused an early birth. I continued to work very hard, and spent lots of time out of town. Judith kept track, and I worked an average of 105 hours every week. It was grueling to all of us.

Our oldest daughter Tracy walked home from school one day, and was frostbitten with only a few minutes out in the cold. Winter activities for most people were snow-mobiling, and ice skating.

Judith was carrying the baby and leading the older two going into a store one day, when she slipped on the ice. She fell flat, and hit the back of her head. I don't think there is anything that hurts worse than landing on your head on the ice! I think that was when the decision was made to move away.

Our last daughter, Holly, was born a year and 5 days after Heather. How did that happen? Judith now had a baby in arms, another in diapers, a 5 year old, and a 7 year old. I was always running off to work, and she felt like she had the world on her shoulders.

I was having real success at my work, which should be no surprise when you realize how much time and effort I spent doing that. The second year of my employment, the company sent me to Ventura California for three months to teach a company school for newly hired Field Engineers. I returned to North Dakota a few days before Christmas, and we tried to put our family back together.

That year we had moved from a rented house, into our own Trailer home, the kids were making new friends, and things seemed to be moving forward. I was able to spend less time at work, and had trainees to help with the work, plus our truck drivers and laborers were requiring less guidance and hand holding.

That following summer we took a vacation, and went to Glacier National Park, in northwest Montana. We camped for a week, enjoyed the beauty of nature, and generally rested a little. It was really good for everyone.

We had a Plymouth car which had only two doors, so the children couldn't open a back door and fall out. They kids were all small, and safety belts were not then installed in cars, so we would fill the back seat floor area with luggage, and made a large play area for the kids on our long trips. They really enjoyed it.

When we drove back to Utah and Idaho to visit our family, it was a drive of about 11 hours, even across Montana which had no speed limits. We only made this trip about once a year because it was so far. It was a circus to watch us go down the road with that car, because we almost always had the trunk open and roped down with so much stuff in it.

We were also able to take a vacation in the Black Hills of South Dakota, at the Mount Rushmore Monument. What an impressive vacation for us all. I will never forget rounding that corner on the mountain road approaching the monument, and seeing the faces of the four presidents carved out of the living rock.

In total we lived in North Dakota for three winters. Judith had two children there, and we were ready for some changes in our life.

Judith was by now very tired indeed, our four daughters were running her ragged. Tracy who was 9 (going on 35) was intent on running not only her own life, but the rest of the family as well. I guess this is normal for 9. Teresa, the pleaser, was 7, and was always "just happy to be there". Heather had just turned two and was truly in her "terrible two's". Holly, the baby, had just turned one, and was as needful of attention as a one year old could be. With everything else that Judith had to do, she was dependent on a husband who averaged over 100 hours per week at work. Considering all these things you get a feeling for what a job she had on her hands.

I often think of Judith's character, and how she could cope with those difficulties. I definitely "married up" when finding her.

In February, 1972, I received word from my company, Schlumberger well services, that I would be transferred from the USA to the Parent Company for reassignment in Dubai. We were slated to go to Dubai, which is in the Trucial States on the Persian Gulf. We studied hard to get a good idea of what we were getting into.

My work, though consuming, was the vehicle which was allowing us to overcome the financial hardships of having a young family, and pay off university loans and do all the things Americans "Need" to do. I had applied for the overseas transfer to give us the added income to pay off the rest of the education loans which my father in law had co-signed.

When our offer of an overseas transfer came through, we were told we were going to Dubai, we were advised that the company would only ship a minimum amount of our family possessions. We had an allowance of three large steamer trunks which would be air-freighted, and our hand luggage which we carried on the flight with us. This was all acceptable, because Dubai was a duty free port, and we could buy anything we needed on arrival. Life was good. So it seemed.

We were living in Williston, North Dakota, a small town of 12,000 populations. We were surrounded by other small towns, and over 120 miles from the largest regional city, which was Minot North Dakota. When we found that we couldn't get our yellow fever vaccinations there, and it required 10 days after receiving them before we could travel to the Middle East, we decided to take what remained of our vacation and go to Salt Lake City, Utah, for the vaccinations. We would then spend 10 days with our families while preparing to go to Dubai.

We stored our household goods at a storage facility, because we were only intending to be in Dubai 18 months, but sold our car. We then took a rental van with the family, and 12 suitcases to travel to Salt Lake City. The date was March 15, 1972. The weather was just turning nice. It was actually above freezing in the daytime, which was a nice change from the January temperatures of 50 degrees below zero. As we left Williston, the Missouri river was beginning its annual spring breakup. Giant blocks of ice more than ten feet on a side would break out of the surface of the river with a tremendous crack of thunder, and jump 40 or 50 feet into the air, only to fall back onto the frozen surface, and break the surface ice for more of the same.

When we got to the Missouri River Bridge, the ice had piled up against the bridge uprights, and there was water beginning to flow across the bridge. The state police were there and told us to either go back to town for another week, or cross NOW. We made our decision, and started across the mile wide bridge. Before we could get all the way across the water deepened to where we had almost a foot of water over the bridge. If we hadn't have had the van's high clearance, we wouldn't have made it.

We crossed safely, although the freezing water did something to the oil pan of the van. After only about 100 miles, the engine began to overheat, and lost oil pressure. We stopped in the first town, and called Hertz. They put us up in a hotel overnight, and gave us a car for the rest of the 900 mile journey. They didn't have another van, but agreed to send our luggage in a separate car to meet us in Salt Lake City. What an auspicious beginning to a trip!

We arrived in Salt Lake City, and got the yellow fever vaccinations, then retrieved our luggage, and spent ten days with our families in Utah and Idaho. We were told 5 days before getting on a flight to Paris, that the office in Paris was unsuccessful in getting Visas due to some problem, and that the only job they had available for us would be in Iran. We had studied all about Dubai, our household goods were on their way to Dubai, and we were expected to go to Iran. The inevitability of the situation finally settled upon us. Schlumberger offered to send our things from Dubai, and send us on a buying trip to Kuwait, which was another duty free port across the Persian Gulf from Iran to purchase our household goods. We finally agreed. We felt we had no real choice in the matter. At this time there was a government ordered freeze on wages in the USA, and this was having a chilling effect on job growth. I didn't have good prospects of finding another job.

We left for Paris on the 25th of March, and arrived on the 26th. We were expecting to be met at the airport by the Schlumberger representative, but since it was a Sunday morning, nobody showed up. After passing immigration control, I tried to call the office to find out where everyone was. No office. You have to love the French. I then placed an overseas call to the Schlumberger office in Houston, Texas, and got the all night emergency duty operator. Still no luck, I then called my local dispatcher in Williston, North Dakota. This lady, Trell Watson was the one who although disabled, handled the phones for Schlumberger from five in the evening until eight the next morning. What a wonderful voice she always had. Finally I got someone who would help. She finally escalated the call to the President of Operations in the USA. In about an hour the Personnel Director for the Paris office showed up at Orly to take care of us. What a relief. We were so relieved.

The French are not what you call a helpful people to others who don't speak French. They may take 10 or 12 years of English in school, but they won't use it except to correct a visitor's bad French. We took three taxis. Two for all the people, and another for all the baggage, and were moved to the Victoria Palace hotel. This was the hotel used by Schlumberger for all their foreign visitors. I have no idea why. It had solid gold cutlery, and a restaurant which I am sure was five stars, but they hated children, and didn't want them in the dining room. By this time a 6'3", 200 pound American who had had enough wasn't in a mood to put up with a bunch of bad behavior any more. That first time lunch however, was a bit "entertaining", only in retrospect. At the time, it seemed an absolute disaster.

We were seated in the dining room. As we looked around we were annoyed at the people staring at us. I guessed they had never seen a family traveling before, maybe the French were unable to have children, and they magically popped up at 18 years of age. Hummmmph. We ordered as best we could from the menu, which was all in French, and I relied on my only foreign language –Spanish- to enable us to order food. They brought onion soup for all. When it arrived, our two year old who was starving by then, having just been weaned from the bottle, immediately began to eat. After two bites she got this funny look on her face, and vomited all over the table. Talk about a fountain! The one year old, who was sitting on her mother's lap, got this funny look on her face, as if to say, I don't like the smell of that either and she also vomited in reply. I guess that is one way of describing how you appreciate someone's wonderful gastronomical delight. Wow what a way to begin a foreign assignment.

I spent two days in the Schlumberger office being briefed on the new assignment, and getting visas put in all our passports. When getting our first passport, we had been told by the company to get a passport for me and another for Judith with all the children in it as well. When we went to apply for the passport, the very helpful passport officer said that while this was permissible, it was not advisable, when traveling to the Middle East. They advised us to get separate passports for everyone, which we did. What a wonderful piece of advice that turned out to be. I never had occasion to regret that advice, and many times was grateful for it. We kept the passport photo for the purpose of work and residence permits.

Judith and girls Passport Photo

During this indoctrination period in Paris, my family stayed in the hotel, and I went back and forth to the Schlumberger office. I still remember that address: 42 Rue St. Dominique. I have always been a bit of an explorer, and I wanted to walk around a little. In case I got lost, I had one of the company secretaries teach me how to say the address in French... She finally said that my rendition wasn't good, but passable enough.

On the second morning of our short sojourn in Paris, I left about an hour early from the hotel and walked by the Palais du Justice, and just had a little time to wander around. I walked and walked, and although I was sure I was near the office, I didn't recognize any landmarks. I finally gave up, and hailed a cab, the surly driver, when I told him where I wanted to go in my very best French; Quarente deux Rue st. Dominique, replied; eehhhh. I then repeated, starting to feel slightly apprehensive, but a little slower; Quarente' deux - Rue St. Dominique. At this point, he made a hand in the air gesture, said, "bon", and drove around the corner to stop in front of the office.

I still wonder if that driver didn't go home and tell his family, "I drove this American today, who was so lazy he wouldn't walk around the corner". Oh well life isn't good unless someone gets a laugh.

On Saturday, March 31, we left the lovely city of Paris (spit on the ground), and changed planes in Marseilles en-route to Tehran, Iran. We arrived on Friday afternoon, which is the Muslim Sabbath. Are you noticing a pattern here? I still remember the Air France pilot when we arrived in Tehran, announcing; Bonjour mesdames et messieurs, welcome to Tehran Iran, the temperature is 85 degrees and you can set your watch back 600 years. It was April 1; in another word April fool's day, and I was having serious considerations that it might have been very appropriate. Our older daughter had a little camera, and when she tried to take a photograph in the Tehran airport a policeman informed her that taking photographs in the airport was considered spying, I guessed that they didn't think anyone else had airports.

We stayed in a hotel overnight, and proceeded to the sunny city of Ahwaz in the south of Iran the next morning.

Ahwaz airport is a small airport, without concourses and jet ways. When we walked down off the stairway onto the tarmac, we thought we had literally gone to hell. The temperature was over 100, and the smell was something out of Dante. I remember my wife's stolid question; "what have you gotten us into"? I was wondering the same thing. It turned out that many families, when presented with Ahwaz, Iran, sat in the airport, waiting for the next flight and were there less than five hours before vacating their employment contract and the country. It was miserable even on April 2. We did have a company representative awaiting us in Ahwaz, and were taken to the hotel, and I was taken eventually to the office.

The Ahwaz hotel was the best oasis in town. In 1972, there was a small foreign community starting to make its presence known in Ahwaz. Until 1972, the oil company headquarter offices in southern Iran was first in Masjed-e Suleiman, which means (King Solomon's Mosque). It was affectionately known as MIS. This little company town is about 100 miles north east of Ahwaz, and was the center of civilization for that region. There were about 3000 foreigners living there who were employees of the oil company. MIS had hotels, swimming pools, parks, company stores and company recreational club areas. All the contractors however were forced to live in Ahwaz. There were no company facilities there at that time for non-oil company members. Everything was on the local economy, and quite primitive. My family was the first American contractor family sent there. We immediately enrolled our children in the Ahwaz international school. I think my two daughters comprised about 10% of the total student body there.

We stayed in the hotel for two weeks, while our house was being finished. My work involved going to the field for several days at a time, and taking electronic measurements in wells deep within the earth. I had a local crew of laborers to drive the truck, and do the lifting etc., but the technology was mine to do. I made several jobs, which only served to verify why I had been sent there. The place was a mess. The company had a French location manager, and the engineering staff were all French bachelors. They didn't take care of their jobs well, and they worked two months in Ahwaz, then a month on leave in their home country. The quality of their work was causing trouble, and the exclusive service contract which Schlumberger was enjoying with the Oil Company was in jeopardy. None of the engineering staff or the manager spoke any of the local language, called Farsi, and their ability to communicate with the laborers was very hampered. When a labor dispute came about, and they were often, there was in interpreter available, but his loyalty was highly questionable. I recognized immediately, that I and my family must be able to communicate freely in the local language, or I couldn't get my job done, and my family couldn't survive.

As soon as we moved into our house, we hired a tutor. He taught English classes at the local university, and came into my home three evenings a week to teach us all Farsi. The children took to the language quickly, like you would expect children to, but I was a little harder to teach. Judith had the hardest time, although she did develop a good control of the language over time. I was even able to read and write the language, which proved to be very, very valuable over time. As my language skills developed, I learned more and more about the problems at the office, and where the sources of trouble really were. It was obvious that the problems were not so much at the engineering level, but at the management level. I would never be able to adequately fix the difficulties at my level, and the manager was so entrenched with his bosses, that I elected not to try to repair that damage, but simply do what was required at the engineering level to try and show a good example.

Schlumberger had a policy of giving a bottle of good imported champagne for each "slick job", which was defined as a trip to the well site where no down time was experienced, no extra runs were required, and no tool failures were experienced. To my knowledge, no slick jobs had been performed until the time of my arrival. I prepared my tools and crew carefully and quickly began being awarded lots of slick job bottles of champagne. Judith and I don't drink alcohol, and so we were building up a quite a store of this expensive champagne. It was becoming a very noticeable thing with the company. As time went on, we developed a circle of Iranian, and Foreign friends. As there was no real cultural drawing place, like theatre, or even anything recreational to do, the professional community had parties. Dinner parties, luncheon parties for the ladies, and card parties were the only viable source of entertainment. We started taking a couple of bottles of this stock of champagne with us to give to the host when we were invited out. We became very popular indeed. I was getting two bottles of champagne every week, and we had a hard time giving it away. Our circle of friends grew quickly, as our reputation as champagne fountain grew. We didn't realize this, but just thought people liked us, until later when the subject was broached at a party. Well, live and learn.

The company finally brought in an English and then another French operations manager. The divide between the nationalities continued to grow, and only became worse with each staff change.

Chapter 2

First Hunt

One of the friends I met due to the habit of entertaining at parties was Mickey Meroney. He was the local manager of Sedco Drilling Company. He and his wife Barbara entertained many of the expats and the local community of professional people as well. The drilling rigs had a weekly delivery of feature films which had been resized to the 16 mm format for use in small theatres. As these films were circulated around the various drilling rigs, Mickey and Barbara also had them in their home. We were often invited to their home to watch a movie.

Our children were very well behaved, and they were accepted openly. It is amazing what a good impression well behaved children make on a host. Because I had a lot of machine shop and gun making experience, Mickey asked me to investigate an antique weapon he had acquired. I knew the gun immediately by its caliber, and design. It was a .577 Snider-Enfield rifle, with one of the earliest breech loading mechanisms.

The Mechanism was retro-fitted to the 1853 pattern Enfield, and was done en masse during the 1860's. The old gun had a breechblock mechanism that flipped over and allowed the user to put a cartridge into the back of the barrel, and then close the breech and fire. 10 to 12 aimed shots per minute were not unreasonable. This was the first really successful breech loading mechanism in black powder weaponry, and was a real history in itself. I was able to tell Mickey about this, and that led to him asking me to go hunting with him. Mickey was well connected due to his position as Vice President and General Manager.

He knew all the Iranian dignitaries, and entertained them often. Because of this he had been able to get help in importing a shotgun and rifle. He asked me to go with him and Bill Whitney, who was the head of the oil company drilling department, who also had a rifle in the country. Believe me; I was happy to accompany these two fellows. On a Thursday evening, we drove out of Ahwaz to one of the Sedco drilling rigs, where we arrived about 5:00 pm.

There was time to take Mickey's rifle and familiarize myself with it, and even shoot it a couple of times. He had a model 70 Winchester in .308 calibers, He didn't know an awful lot about guns, and when I shot it the first time, I found that it was not shooting true to the scope position. I told Mickey about this and he asked me to adjust it, because he didn't really know how to. He had wondered why he couldn't hit anything with the gun, and now he knew why.

I adjusted it and it shot wonderfully. We then went back to the rig to have dinner at the camp kitchen, and wait until it got dark when the hogs would come out to eat the crops. During dinner we talked about hunting, guns, and hogs. I asked Mickey and Bill where they should be shot, and they replied that it would be just fine to shoot them right behind the ear, as that would kill them instantly, and make them easy to find by flashlight. In Iran, hogs are considered untouchable by the local people and vermin by the law. You can shoot all you want, and in fact are encouraged to do so.

Hogs will decimate 40 acres of cucumbers, watermelons, or spring wheat in a single evening, so anyone hunting them is extremely welcome with the farmers. A big hog with sharp tusks which may go to 18 inches long is a very dangerous animal. Every village has tales of farmers who have fallen prey to their slashing wounds, and have lost a limb, or even their life at the hands of these creatures. A big hog may weigh well over 1000 pounds, and are really mean and nasty. After dinner, we took the pickup truck, and some spotlights which plugged into the cigarette lighter of the truck. In those days good halogen spotlights were not available, and we made them up from headlights. They were good for about 100 yards, but quickly faded out after that distance.

As a guest, it was suggested I took first shot. I asked if follow on shots were permissible because the hogs were often encountered in a herd. Mickey and Bill both said that I was supposed to shoot until I was out of cartridges or satisfied. They were having fun with me, but I didn't know it. We weren't out of the rig camp for about 15 minutes before we saw the first herd of about 100 hogs. We stopped and since it was pitch dark, except where the light shone on them, I asked Mickey to hold the light on the background.

I didn't want any stray bullets addressed "to whom it may concern". When we were certain the backstop was clear, Mickey held the light on the hogs which were about 125 yards from the road. I quickly lined up and fired. The hogs began to run. As they ran, I picked out another and fired again, then repeated this until the gun was empty. Four shots total. Mickey and Bill asked me if I had hit anything, and I said I thought that I had. We then reloaded the rifle, then took shotguns and flashlights and went out to where I first shot. There where I expected lay the first hog. As we looked at this hog, Mickey gasped, in incredulity "you shot it right behind the ear". I finally got the picture. I then said "well, we'd better go check the rest of them". More blank stares.

There were three more hogs lying within 100 yards of each other, each shot while running – right behind the ear. My father was a superb rifleman, and since we were poor, he raised me not to waste ammunition. The story was told and retold within the expatriate community in Ahwaz. More importantly, I had four hogs to take back to Ahwaz, to be used in lots of entertainment dinners for Mickey, Bill and for Judith and me to supplement the stock of champagne which we had at our dinner parties. My reputation spread as someone remarkable. Mickey and Bill thought this was a superior feat of marksmanship, but in my Father's eyes, it would have been expected.

We built a swimming pool in Mickey's back yard that summer, and it had a big barbecue rotisserie. Any time he had a party there was a fresh haunch of roast pork. These haunches would run about 50 pounds after being deboned. What a wonderful treat, when you couldn't buy pork in the market. The hogs were so much cleaner than the cattle were. Local beef was the meat which must be cooked thoroughly, because of the dirty conditions they were raised in. Beef there was full of trichinosis. The only beef which was safe was the wild water buffalo, or imported.

Early that next summer I decided to leave the employment of Schlumberger Well Services. My job was going nowhere, because of the management attitude, and I was fed up with it. I approached the oil company, and was told that because of the contract they had with Schlumberger, they couldn't even talk to me about a job while I was still employed *by Schlumberger, in* Iran. I was also given a "wink and a nod" and told that they would be happy to talk with me otherwise.

I went ahead and resigned my job with an effective date of September. We left our personal things all except our clothes, all packed up in shipping cases, and our pet dog which we had acquired in Iran with our Canadian friends, Conrad and Lillian Eliason, and we went back to the USA.

When I resigned, the Schlumberger management in Iran asked me what it would take for me to stay, and I told them my salary requirements, and that I would only stay as the director of the training facility in Kuwait. I remember the manager laughing at my requirements, saying that I could never get that kind of money. With a little smile I later faxed him a copy of my first check with my new employer. Sometimes what goes around comes around.he real Iran is not found in its cities. As in any other predominantly rural society, the city is a small measure of the character of the people. The mountain and desert nomads hold the key to understanding the people of this great land. Their traditions lie steeped in over 3500 years of history, and by understanding them we can grow towards understanding the cultural roots and pulse of the modern Iranians. I look forward to the day when I may again sit together with the rural people, and laugh, or talk about the inconsistencies of life, without the specter of governmental conflict hanging over us.

Beauty in the Desert

Chapter 3

The Meeting

"I would like to speak with Mr. Masoud Akhbari", I told the secretary. She motioned for me to follow her, and she stepped into the office. It was the month of June and the temperature outside was over 120 degrees Fahrenheit in the shade, and the air-conditioning systems in all the offices were running day and night.

I followed, and saw a pleasant man sitting behind his desk which was covered with papers. He looked very busy, and had three telephones on his desk. He had one in each hand, pressed to both ears simultaneously, and the other one was ringing incessantly. He quickly finished his two phone calls, asked his secretary to take care of the other line, and greeted me warmly. I was embarrassed to call on this gentleman, whom I didn't know, without any appointment, but time was short, and I needed his advice.

After a few moments of formal greeting as is required in the Iranian culture, he enquired what he could do to help me. I said, "I work for the Schlumberger Company, and Mr. Mickey Meroney has suggested I speak with you".

"I have lived here for 18 months, and I am leaving to go back to the United States", I began. "I expect to receive an offer of employment from the Oil Company to return, once I am back home", I continued, "but they can't offer me a job here while I am in Iran. If we can negotiate terms, I will return to Iran in about 3 months, and I would like to bring my guns for the purpose of hunting. I was told by Mickey Meroney, that you were the best hunter in the area, and had knowledge of how I could proceed."

"Please sit down Mr. Taylor," he said. I immediately took a liking to this man. He was open and forthright, with dark brown eyes, and dark hair. He was quite tall for an Iranian, standing about 5 foot 10 inches, he reminded me of my father with his Mediterranean complexion and open face. "I am embarrassed to come unannounced", I said, "but time is short, and if I can't bring my guns back and go hunting, I don't think I am interested in working here anymore".

He immediately set me at ease, inferring that there was nothing more important to him than talking to me for this very minute. The kindness of this gentleman has grown on me over the years; I have never seen him show an unkind action to anyone in almost 40 years that I have known him. He is without a doubt the kindest man that I have ever known.

"It is very difficult to bring guns into Iran," he continued. "The government has strict regulations on shotguns, and it is almost impossible to bring a rifle into the country, unless you are military". Suddenly I felt disappointed, here Persians say "Abe pak roye dastam rikht" which literally means, "he poured clean water on my hands"; (to let someone know help is not possible), and I guess it showed in my face. He then encouraged me, and said that if I really wanted to go hunting that badly, then he could make it happen.

In Iran there is a complex and most inclusive set of manners which everyone is expected to observe. This is called ***Taarof***. I had experienced ***Taarof***, and as often as not, it resulted in beautiful, flowery conversations, that led nowhere. As an American, I had been raised in a society that respected productivity over all else. I was suspicious of a society that emphasized form over function to the degree that I saw in Iran. I'll give you an example of ***Taarof***.

If a person is at a big party, and finds themselves with their back to someone, they would recognize their faux pas by saying "I am sorry to turn my back on you, please forgive me". The other person would reply, ***"Gul Pusht va Ru Nadareh"*** means "A flower has no front or back". In the Persian language and culture, ***taarof*** demands that you should save "the water of one's face" means ***(Aberoyash ra hefz kon)***, by minimizing your own feelings and reactions, while concentrating on another's needs. ***Taarof*** is widely recognized, and in fact marks the difference between cultured and uncultured people in Iran.

For example, if you ask someone to do a favor or a job of work for you and he is willing to do it he replies, ***"Roye Cheshmam"*** (on my eyes), the answer to this should be ***"Cheshme shoma roushan"*** your eyes are always bright. In Iran, when someone says something nice to you, you have to be careful not to take it too literally. It may just be Taarof. It is not considered to be a lie to tell someone something wholly untrue, if it will make them feel better. I was made rather painfully aware of this on several occasions.

It was a practice for instance, when asking road directions, to ask three times for the directions from the same person. On the first two occasions, you might get nonsense, because the person doesn't want to disappoint you by telling you he doesn't know, but the third time, they will tell you correctly. The previous 18 months experience with Persians warned me to be careful, but my instincts told me to believe this man. He went on to give me explicit instructions on how to leave the guns with customs when they came back to Iran, and all the details I needed to know.

Here is an example of Persian Colloquialisms for enjoyment and better understanding of the Persian mentality:

This letter was written by an employee of the NIOC (National Iranian Oil Company), Ali Mokhles, back in the 1960`s to his American boss, Mr.Hamilton. The superscripted sayings are explained in the number indexed text below.

Dear Mr.Hamilton:

I, the undersigned, have worked in the NIOC in Masjed-Solyeman for three years, But since Mr.Ahmadi transferred here everything has changed. I don't know [1]"what a wet wood I have sold him" that from the very first day he has been [2]"pulling the belt to my left" With all kinds of [3]"cat dancing" he has tried to become the [4]"eye and the light" of Mr.Wilson.

He made so much [5]"mouse running" that finally Mr.Wilson [6]"became donkey", and appointed Mr.Ahmadi as his right hand man, and told me to work [7]"under his hand" Mr.Wilson promised me that next year he would make me his right hand man, but [8]"my eye did not see water", and I knew that all these were [9]"hat play", and he was trying to put a [10]"hat on my head" I [11]"put the seal of silence to my lips" and did not say anything.

Since that he was just [12]"putting watermelon under my arms" Knowing that this transfer was only [13]"good for his aunt", I started begging him to forget that I ever came to see him and forget my visit altogether.

I said [14]"you saw camel, you did not see camel"....but he was not [15]"getting off the devils donkey" ... [16]"what headache shall I give you" I am now forced to work in the mail house with bunch of [17]"blind, bald, height and half height" people. [18]"Imagine how much my rectum burns."

Now Mr.Hamilton, [19]"I turn around your head" you are my only hope and my [20]"back and shelter"....[21]"I swear you to the 14 innocents" please [22]"do some work for me"....[23]"I`ll grasp your skirt"....[24]"I have six head bread eaters".....[25]"I kiss your hand and legs".

Your servant,

Ali Mokhles

Mr. Hamilton replied: [26]"My liver is barbecued"

Here are the meanings of the Idioms:

1 - "What a wet wood I have sold him": He thought I had sold him a "bill of goods"

2 - "pulling the belt to my left": An old Persian Proverb meaning that when helping someone up by grabbing their belt, pull from the center, otherwise you may do more harm than good.

3 - "Cat Dancing": Think of a cat dancing on it's back legs. It will do a lot of jumping around the subject

4 - "Eye and the Light": Going blindly after what someone says or does; depending on someone to an extreme.

5 - "Mouse Running": Going around in circles

6 - "Became donkey": Stupid and obstinate

7 - "Under his hand": Close subordinate sharing a good relationship

8 - "My eye didn't see water": I didn't believe what was told or promised to me

9 - "Hat play": Not treating someone honestly; kind of like the old western "shell game" with walnut shells and a pea

10 - "Putting a hat on my head": Taking advantage again after burning me once

11 - "Put the seal of silence on my lips": Since someone has taken advantage of me, I won't talk with them anymore

12 - "Putting watermelon under my arms": Picture a person with a big watermelon under each arm. Their arms will stick out making them look bigger than they really are. The proverb deals with making a person look more important than they are

13 - "Good for his Aunt": Empty promises

14 - "You saw camel - you did not see camel": I came to you and talked to you but you acted like I didn't talk to you, we would say “on again -off again”

15 - "Getting off the devil's donkey": You're making a mistake that will lead you to ruin

16 - "What headache shall I give you": I don't want to give you a headache with my problem

17 - "With blind, bald, short, and tall people": All sorts of people, generally derisive

18 - "Imagine how much my rectum burns": Hard to translate. Generally it means angry, but more. Really and constantly irritated

19 - "I turn around your head": When people go to Mecca, they turn around the Kabba Stone. It is a term meaning I honor or respect you.

20 - "You are my back and shelter": You are the roof over my head; again a term of respect

21 - "I swear you to the 14 innocents": This refers to the 14 IMAMS, invoking a spiritual request.

22 - "Can you do some work for me": I have something for you to do, or Can you do something for me?

23 - "I'll grasp your skirt": I am asking you to do something, and holding onto your clothes to emphasize my need; I won't give up easily with my entreaty

24 - "I have six head bread eaters": Six in my family

25 - "I kiss your hand and your legs": Obeisance, flowery way of saying respect

26 – " My liver is barbecued": This means that I would like to do something, but I am frustrated at my inability to help.

Mr. Akhbari, (the name means great one) told me to call him by his given name, Masoud, and a strange kinship immediately developed between he and I. At that moment I felt, as Persian proverb says, "***Del be del rah dare***", meaning both persons perceived their friendship and sincerity. I have always been a loner, not prone to making quick relationships, but I had the distinct impression that I had travelled half way round the world to find my brother.

I recognized in his demeanor that he felt in a similar way. We discussed hunting for a few minutes, and I tried to get some idea of what kinds of things he would be in need of from the United States. Iran had about 400% duty on luxury goods. Many things cannot be imported at all. These not-to-be-imported things were most things that had anything to do with guns or hunting. Coming from the United States, it was strange for me to have such strict gun control, but that is the way it was in their country, with their laws and customs.

During the next three months in the United States, I put together everything my family would need for the next year, and all the hunting and camping supplies I thought we would need. I went to the mountains of southern Utah for a deer hunt, then back for more negotiations and job details. After a month of negotiation, I accepted a job offer with The Oil Service Company of Iran (OSCO), to do petrophysical evaluation and well-site quality control.

The contract was in two year increments, for as long as I wanted to work there. I had paid off all the school loans and medical bills while on my first 18 month assignment, and was looking forward to saving some real money.

I had a hand built Mauser 25-06 rifle, with a fine German scope made by Pecar Berlin Company. This extremely accurate rifle would keep 10 shots within ½ inch at 100 yards, if I did my job. I took it, along with my "old reliable" 12 Gauge Remington pump shotgun, and my family and I got on the airplane to go back to Iran.

This time things would be different; we weren't working for a private company that had little or no infrastructure in Iran, we were working for the "Oil Company". They had a wonderful host of amenities for employees. There was a health Clinic, hospital, commissary, swimming pool and movie complex, and then the golf club and softball club were there for us to join if we wanted. We were very much looking forward to the change of conditions which were to be found with this new assignment. My children were 9, 7, almost 4, and almost 3. They were excited to see their old friends, and to have a new house and carry on the adventure.

My wife, Judith, bless her heart, was just tired, and wanted to rest from the hassle and transatlantic moves. I had worked long hours away from home for the past 5 years, and she was looking forward to having a husband to help with the difficulties of raising four lovely but energetic daughters. I too was looking forward to having some time at home. Little did we know how hectic our life was about to become. We arrived in Iran in late October, 1973. We had brought a lot of things for us and the children, which we didn't have during our first assignment. We had an excess luggage allowance, which was full to limitation, and we had an air freight shipment on its way too.

We knew we would have some hassle with the customs when we arrived. Iranian customs agents are not paid well, and they are always seeking a "gratuity", to expedite your trip through customs. It seems that many things that may be in luggage are not prohibited, but not explicitly permitted either. This gray area is fruitful ground for negotiation of just how much "gratuity" is required to clear these things.

We were in for the shock of our lives. Never did I ever suspect what a Godsend these guns would be for my family. When we arrived at Tehran Mehrabad airport, in the customs hall, we got a big flat-bed trolley, and collected all our luggage and excess luggage on it. There was this ominous piece of luggage which to everyone that knew anything about guns, simply spelled "GUNS". I took it to the "something to declare" desk along with all my papers as I had been previously instructed. My wife engaged a laborer to take the trolley and our four children out of the customs hall. We were expecting something like a couple of hours to get all these goods through customs. When I opened the gun case literally every customs agent in the airport zoomed in on me like bees to honey. Everyone had to speak to me, and hold these wonderful, rare things.

They pointed them, played like they were shooting and hunting, and generally displayed what a rarity they had found. In the meantime, my family, and every other family which had any wits made quick their escape from the customs hall. They simply took trolleys and walked out to the waiting taxi ranks. I stood and looked forlorn, wondering how long it would take me to get this mess cleared up. Within minutes, a supervisor realized how unmanned the place was, and ordered everyone back to work. They locked my guns in their case, took them away, and gave me the required receipts. I walked out of Mehrabad airport customs in 15 minutes, a new international record, and not a single "gratuity" had been extracted from me! When I met my family outside, there was a waiting cadre of people waiting to cheer, because they too had the good fortune of clearing customs without hassle.

We commented while waiting to get a domestic flight to continue our journey to Ahwaz in Southern Iran, just how many schemes there were to "fast and easy Iranian customs clearing". Those schemes were communicated among the many foreigners in the country, and included; a light pinch of a very cranky 1 year old; (Iranians hate to see children cry, and would do anything to speed a crying child and their family on their journey). Another popular scheme was feigning illness which sometimes created enough of a diversion to allow a hapless traveler to escape from the customs hall. (The problem with this particular scheme was that the perpetrator might have to be examined by a doctor if they didn't play it just right). The "guns in the case", diversion was the best anyone had ever heard about, and because I had to take the guns out of the country every summer when we went on vacation, we developed a reputation. People in the international community would press us for our travel itinerary months before summer vacation, and arrange to be on the same flight with us. We always accepted this new found popularity with some degree of suspicion that people really didn't like to fly with us that well, but we never were really sure....

When we arrived in Ahwaz, it was actually cool. What a pleasant surprise. When we left three months earlier, it was over 120 degrees Fahrenheit. Now it was down to a mere 85 degrees in the early afternoon. We expected snow any minute. What a change from the first experience. We only had to stay in the hotel a few days, as opposed to almost a month. We had a brand new house to move into, and our air freight shipment was received only a few days later. The children were quickly integrated back into the international school, and Judith immediately began to make our house a home.

I took my receipts, and papers to my new found friend Masoud and he began to work his magic. I had learned to speak passable Farsi (the Persian language), during our first assignment, but the word set which is used for working and shopping are very different than those words used in conversing with very high ranking diplomats, and government officials. I had to learn a new language quickly. I spent considerable time with Masoud being tutored in the culture, and the polite way to act and say things to these people.

I remember one occasion, where I needed to give a verbal deposition to a member of the secret police. His name was Sarhang (Colonel) Sharokhi. Masoud called him on the telephone, and then handed the phone to me. I was expected to carry on a meaningful conversation with this man, and convince him that I wasn't a criminal, and that I would always act with consideration, and safety. Moreover, I had to convince him that he wouldn't be personally embarrassed by any misbehavior on my part. It was a grueling time.

I should explain that most Americans in Iran couldn't string a dozen words together. They didn't need to. They had house servants, who did the shopping. They had guards to guard their street. Many had drivers to drive their cars. I was the odd one because I could speak passable Farsi. That meager command of the language, however, paid me many times over its investment in time. I learned that if a person makes an effort to speak in the native tongue of a country they are visiting, even though they are not fluent, it is really appreciated by the local inhabitants. They will do far more to help you if you try to help yourself.

When my conversation was finished with Colonel Sharokhi, I handed the phone back to Masoud. When he finished the conversation, I learned from hearing what he said that he was personally guaranteeing my conduct. I was so impressed. I barely knew the man, and he was going way out on a limb on my behalf. I didn't know what I had done to earn such trust. Such has been our relationship ever since. I am as I have explained an active Mormon (Church of Jesus Christ of Latter-day Saints), and Masoud is a Muslim. Those religious differences mean nothing to us. I feel it would literally be an honor to give my life for this wonderful man or one of his family and I am confident that he feels the same way about me and my family.

A few days after my talk to the Colonel Sharokhi Masoud came to my office and said that the Colonel had invited us for two weeks hunting in his village at the foot of the Zagros Mountains on the Esfahan side. This was planned for 18 months later! We were informed that it would take some time to get my permits. The next step was to wait. The month went quickly. I had a new job and responsibilities to settle into, a new house which required the minor modifications which would make it our home, and many phone calls with Masoud to discuss hunting. While we were waiting for my gun permits, I also applied for hunting and fishing licenses. These licenses are not the walk in and buy over the counter kind of thing that we are used to here in the United States. No you have to fill out applications, submit lists of character witnesses, and basically jump through hoops. The day finally came that I had the licenses, and the permits came from Tehran. Now I needed to go to Tehran and collect my guns. I travelled on the Oil Company plane to Tehran, collected my guns, and returned to Ahwaz. I still remember when I first showed my guns to Masoud. He had several nice shotguns and rifles, but he was very impressed with my rifle. He couldn't believe that I had hand built it myself. It had a rosewood stock, with ivory spacers under the recoil pad, grip cap, and fore end tip. It truly was a thing of beauty. When we took it out to shoot it he was even more impressed. This rifle shot like a dream. The recoil and muzzle blast was minimal, and the accuracy was phenomenal. It was accurate and deadly at over a half mile, as was proved on many occasions.

I had brought with me supplies for reloading all of our rifle ammunition. This was of tremendous value. Not only was ammunition very expensive, (about $5 per shot), but it was extremely hard to get. Importation required permits, and only 100 rounds could be imported every two years. By bringing enough reloading supplies to supply over 2000 rounds, I had guaranteed several years of hunting. Rifle cartridges for hunting were not manufactured in Iran, only military calibers. Moreover it was not allowed to have a hunting rifle chambered for the military calibers. This served to control the private ownership of firearms and ammunition very well.

The Friendship Develops

Over a very short time, our friendship rapidly developed into a brotherhood. On many occasions, Masoud would stop at my home, pick up my rifle, and hunting gear, then stop by my office at the summertime closing time of 2:00 pm, and off we would go to the mountains nearby for wild sheep and Ibex hunting. This occasioned many hours in each other's company. I learned a lot about Masoud, and he learned a lot about me. I said that I had moved half way round the world to meet my long lost brother. A brother of a different culture and religion, but a man with whom I have a mutual trust of the highest level. I certainly would trust him with my life and have done so many times. I was able to extract this information from him about his life and background.

Chapter 4

Masoud Akhbari

I was born on the 21st July 1931 in Masjed-i-Sulaiman; (M.I.S) in the Male Karim area of Southwest Iran; the literary English translation is *"King Solomon's Mosque"*

It is a habit of people to have a short name for places, and the name M.I.S was adopted by oil company members and employees, and quickly adopted as official. Even now nobody calls it Masjed-i-Sulaiman any longer.

My father and his brother immigrated to India in 1905. He studied English and after his graduation he was invited by his class friends to take a role as an actor in Indian silent movies. He played in two silent films named "The Queen of the Jungle" and "The King and the horseman", which were in all Indian cinemas for more than 4 years!

He soon left the cinema world and joined the Persian legation in Bombay where he worked as Charge d'affaires for the Persian Embassy in Bombay. There he came into contact with my mother's father, a diplomat who had been put under house arrest and exile in Bombay by the British government during their invasion of the south of Persia in 1914 -1918. My grandfather was arrested in Bushire, which is a port city on the Persian Gulf, and sent under house arrest, by a British naval ship to India. Later on at his request the British authorities agreed to let his family join him.

After my grandfather was released and permitted to return to Persia in 1924 he was appointed by the Persian Government as Governor of Mohammareh; the name was changed to Khorramshahr during the reign of Reza Shah, the Ruler of Persia from 1926 to 1943.

During his stay in India he came into contact with my father in the Persian Embassy. While in Khorramshahr, my Grandfather wrote to my father urging him to come back from India, and work for the Anglo Persian Oil Company *(APOC*), because they needed people with a good knowledge of the English and Persian languages. My father returned in 1925 and immediately he joined *APOC*, founder of the first oil fields in M.I.S., and in fact in the entire Middle East. The first discovery was made on the 23rd of May 1908 in M.I.S. My father married my mother the same year and moved with her to M.I.S. and lived there until his retirement in 1961.

My School Years

One of the things Reza Shah did in his reign between 1926 and 1941 was to ban the wearing of the veil or scarf (Chador or Rosary) for women who were observing Islamic tradition. One of my earliest memories was when I was walking with my mother in the market of MIS. She was wearing a scarf, and the police came and took us to the police station where they removed her scarf and tore it up, sending us on our way and telling my mother not to wear it anymore.

My father was head of the narrow gauge railway between MIS and Dar-i-Khazineh which was a distance of about 60 miles. There were 12 stations along this route, with a place for the engine to take on water, freight, animals and a few passengers at each station. The trip of 60 miles took about 8 to 10 hours for transit because of the route through the mountains and the time which it took to stop at each station.

I remember a funny incident when I was about 14. I was in the habit of practicing driving one of the locomotives engines around, moving carriages along the tracks very slowly, and because of my father's position of authority, no-one dared to tell me not to do this. One day I got going too fast, and then I panicked and threw on more steam, instead of applying the brake. I broke down the gate to the train yard and depot, and went about a half mile before I could stop. I quickly brought the engine back to the depot where it belonged, and put it away. I then jumped on a horse, and rode quickly to the next village. When my father saw the damage, he didn't know who did it, and since nobody witnessed the accident, he didn't know who to blame. It wasn't until after my father's retirement that I told him about the incident. After I told him, he said, "I knew it was you that did that, because you rode away on the horse to the next village, and there wasn't any reason for you to disappear"

During the period we lived in Dar-i-Khazineh, (DIK), my father could run the Oil Company railway and the stern paddle boat efficiently. I went to school in MIS one week, and then back to DIK for the next week and there were two days per week that the MIS school provided a teacher to come to DIK to teach us.
The largest city in the area was Ahwaz. Several times each year we would travel from DIK to Ahwaz by paddle steam boat. It would take about 24 hours to travel this 55 km. We would have a rest stop at Band-i-Qil (BIQ), for about 4 hours. BIQ is near the confluence of Karkheh and Karun rivers. Because the oil company ran all the transportation facilities in this area, the paddle steamer was free for our use, and we could travel as often as we needed without cost.

My father used to hunt often in Persia. When I was 12, he asked me for the first time to join him, and I was happy to carry his back pack or sometimes I was allowed to carry his long barreled muzzle loading rifle even though it was a rather heavy gun. I often begged my father to take his gun back within an hour's duration.

My First Job with the Oil Company

I joined the Oil Company in 1949 as a learner clerk, working in the Geological and Paleontological department in M-I-S., and after a few years I was promoted to Cartographic Draftsman. For almost three years after the Oil Company was nationalized in Iran the work was at a standstill, even though we used to go to the office every day there was no work for us. We sat there counting hours to go home and hoping that someday that we would be busy again.

After the political situation stabilized and the experienced staff of oil consortium (IOE&PC) returned to Iran, the drawing office became active and more and more staff joined in, and after a while we had more than 20 personnel in our department. In 1962-63 I was sent abroad by the Oil Company for training in aerial survey and photo geology to Delft University in Holland and followed by studies at the Ordinance Survey Southampton in the UK.

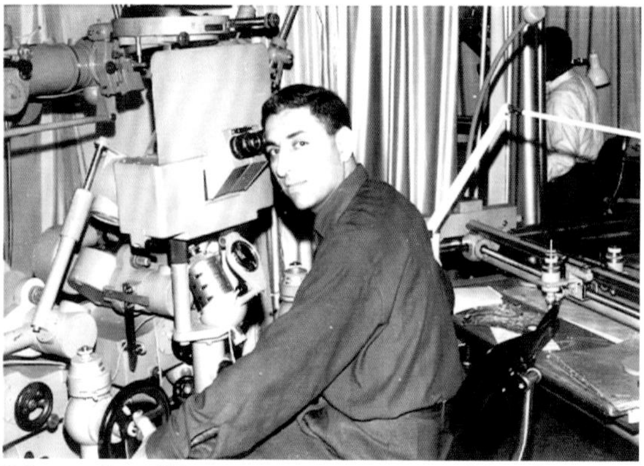

Masoud at Photogrammetry machine in Delft

In Delft I stayed in a house on ***Zuidwal no.****7* close by the Institute and there was an Iranian gentleman who had a scholarship from the Iranian government studying there and living in the dormitory. One day he asked me to join him to go to The Hague to meet two Iranians who had come from MIS for a short assignment like me. I had seen them daily and knew their names while I worked in MIS. The appointment was on Friday 18th January 1963 at 17:00 hours in a restaurant called Piccadilly in The Hague. As we entered the restaurant we saw there were two beautiful girls sitting next to our table, one with open eyes and lots of expression who keep looking at me, and I was also not able to keep my eyes away from her.

Masoud and Willie in the early years

Finally I dared to ask her for a dance, and after a courtship, we returned to Iran, and we were married in Tehran, Iran on 22nd November 1963, the day that U.S. President John Kennedy was assassinated.
In 1969 after 20 years working for the Oil industry I resigned to work as a Consultant to the Oil Company for drawing and reproduction services and the engineering office, for the next four years. During this time, I acquired an iron oxide mining company in the north of Iran, and after my contract was over with the Oil Company, because the demand for iron oxide was increasing rapidly, I decided to go and take care of my Iron Oxide Mining Company. Just as I was about to leave Ahwaz I was called by Bob Ramsey, Manager of the Reading & Bates Drilling Company asking if I would be interested to work for him as area personnel manager.

I accepted, because of the opportunity. It was there that I met my American brother Blaine Taylor.

Masoud (kneeling right) and a wild boar taken in his younger years.

Chapter 5

History and Background

In Persia, and in general across the entire Middle East, the school system was run by religious clerics (Mullahs). It was normal that the students sat on the floor, on carpets, and studied mostly very basic subjects, but most of the lessons were based on Quranic study; the Koran. The schools are called Madreseh' (School). In Persia, because of the influences of Nasser-al-Din Shah, after his various visits to Europe, the schools became more secular.

More scientific subjects were being taught, and in the school in M.I.S., it was sponsored by the Anglo-Iranian Oil Company, and was therefore even more secular than the other parts of the country. For example the sports which were normal in the United Kingdom; Rugby, Soccer Football, Table Tennis, Boxing, and Basketball, etc. were all taught in the M.I.S. schools, and moreover the A.P.O.C. provided the facilities in the same way as they did to the U.K. Schools. The similarity to English schools didn't stop there, however, because of the sponsorship and money available, the scientific studies and English studies were much advanced.

Better teachers were employed, and there were even laboratories for the students to use. The English teacher was an English lady, brought in by the A.P.O.C. just to teach English. The APOC didn't do all these things entirely out of the goodness of their heart, but from their desire to have more qualified local staff, instead of having to bring all of their staff from abroad. This proved to be of tremendous value, for the A.P.O.C., and for the students as well. Many of my class mates were engaged by the oil company, and in fact formed the nucleus of the Persian side of the oil company. We were more like a family than a business concern. This trend continued until nationalization.

There were no universities in the country, but Nasser-al-Din Shah encouraged many students to attend higher education abroad. This continued until after 1932 when Reza Shah started the first university in Tehran. Slowly, more and more universities were established until today most major metropolitan areas have a university.

I grew up in M-I-S., and because I finished my primary and secondary school studies there, I have a much higher quality education than most Persian students of my generation. For further higher education I attended schools in Ahwaz which was a larger city and was the capital of the Khuzestan province, about 100 miles south of M.I.S. I found natural science the easiest and most enjoyable of all my studies. I also studied music on my own, and played both alto and tenor saxophone, but due to concentration on my work, I stopped playing.

The oil company provided three different clubs for different classes of employment; working class club, junior staff club, and senior staff club, the last of which included most of the expatriates. The junior and senior staff clubs were very much alike in facilities, but didn't allow much social interaction between classes. They provided things like billiards, badminton, table tennis, basketball, tennis, and cinema. The cinema was all in English, with no subtitles. Until the end of World War II the cinema began with the British National Anthem, (God Save the King). All must be in attendance had to stand quietly at attention until the anthem was finished, or have their attendance terminated.
The oil company provided a free bus all over MIS. Three different classes of bus were provided for use of the three different classes of employment. This continued until after World War II. On September16th 1941, when the British exiled Reza Shah to Mauritius, they appointed his son, Mohammed Reza to replace him. He was forced to leave the country for about 2 months in 1953, but was restored to power with the help of the super powers. He delayed his coronation until 1967, and then was crowned.

Iran's tumult in 1907 thru 1909

The Bakhtyari history goes back to nearly 480 BC when Herodotus in his history collections refer to the Cavalry soldiers from Zagros steppe. He says that they provided major forces to Cyrus the Great and King Darius in many occasions to fight against the Greeks, I wonder if they were not Bakhtyari tribesmen.

During campaigns with the Greeks Bakhtyari Sowars did not wear the same uniforms as the regular Persian soldiers in service of the Shah, that is why the Greeks named them Barbers. We do not have much information about Bakhtyari Tribes before 14th century and also after that no one put on records any major history of this nation.

Very little is written about the Bakhtyari peoples until the Bakhtyari Khans started to unite their clans of Chahar Lang and Haft Lang In the late 19th century. Only then did one of the Khans start to write a brief history of their ancestors. The Bakhtyaris also helped Safavid kings to fight against their enemies.

After Shah Abbas, we know that Nader Shah also used soldiers from the Bakhtyari tribes during his campaign against India. During the Assyrians era the

Bakhtyari warlords made brave stands against the Assyrians during their attempts to invade the Zagros Mountain regions.

The Assyrians were desert fighters while the Bakhtyaris were mountain warriors. The Bakhtyaris, because of their limited numbers, could only leave disorder and terror among the Assyrians in such a way that they had to abandon their campaign against the Zagros inhabitants for many centuries to follow.

Bakhtyari tribes also controlled the Khorasan Highway which joined to the famous Silk Road and provided protection to the travellers and merchants. One of the famous stopping places of the silk road was in Esfahan, in the caravansari there, now called the Hotel Shah Abbas.

Over several centuries these great warriors of the Zagros were the best horse breeders and all neighboring kingdoms and groups secured their horses from the Bakhtyari nomads.

In about 450 B.C. when Median king Cyaxares led his expedition against the Assyrians, the Bakhtyari Cavalry played important role in conquering Nineveh the second capital city of Assyrians after Babylon.

The Bakhtyari confederations were major land owners in Chahar Mahal va Bakhtyari and Khuzestan to the south-west of the Zagros Mountains. The central government in Persia was always alert and fearful of a general uprising of the tribes, especially the Bakhtyaris.

For this reason during the reign of Nasser Al-Din Shah they granted Husain Qoli Khan and Hadji Ilkhani lots of land and they were exempted from paying taxes.

Ordinary Persian people have merely names, but it was common for them to adopt a title. For example some would adopt the title of "Marshal of the Marshals" or "Unique one of the Persian Kingdom" or "Fortune of the State" they used to obtain some form of parchment to certify his title ownership. I can imagine that it would be rather difficult for foreigners to remember these appellations.

Muzaffar Al-Din Shah Qajar died on January 4th, 1907 but few days before his death, he had signed and granted people a code of laws or constitution rights. This was caused by an almost bloodless revolution, by the Persian people. The centuries old absolutism of the kings in Persia had been legally changed and modified by these writings.

This declaration could be compared to the British Magna Carta, in its importance to the Persian peoples.

The Persian congress, or Majlis had to approve the modification, the first Majlis was opened and the speech of the Shah was read, at the death of Muzaffar Al-Din.

On January 7th 1907 Muzaffar Al-Din Shah Qajar was succeeded by his son Prince Muhammad Ali Mirza, governor of Tabriz. He arrived in Tehran on and was immediately crowned on January 19th, 1907. It is thought that Muhammad Ali hated his subjects from the beginning of his succession to the throne. He quickly began to ignore the Majlis and ordered no open discussions to be allowed.

He aligned withwith Russian emissaries against the Persian citizens, and made contact with England and Russians to arrange for a secret loan of half million British Pounds. This was soon discovered by the Mullahs, Constitutionalists and Majlis, and the deputies determined to assert their power to confirm the reform which was urgently needed. They immediately approved measures to stop any further loans from England or Russia because they regarded fast increasing debts as a slavery of the country.

The Shah had to call on the famous Aminu's-Sultan. He gained fame under the title Atabak-e Azam, or prime minister for Nasser al Din Shah, but was exiled. He was a strict monarchist, and didn't' believe in power of the people. Upon his arrival in Tehran on the 26th April he found a state of disorder and chaos in all departments of the government. The treasury was empty, and there were disturbances throughout the whole country.

The Shah stopped the Majlis from continue their normal work, and ruled that they must do nothing except to carry out his plans.

In Esfahan, the people protested against the rule of Zille Sultan. The city of Tabriz was in foment, and matters went from bad to worse. During August, Russia, which had never been happy with establishment of the constitution regime in Iran, began to threaten the members of the Majlis with intervention.

Troubles with Turkey also reared its ugly head when 6000 Turkish soldiers entered into Iran and occupied some villages along the Turku-Persian frontiers and were advancing towards Uromieh.
The newly installed Prime Minister, Atabak was all this time working hard to obtain another Russian loan, but he was very much afraid of signing any contract with Russia without the approval of the Majlis. At the end of August he

had nearly succeeded to get most of the Majlis deputies on his side to approve his plans. On 31 August he was assassinated, by a young man who was a member of the secret political society for saving the constitutional government.

With all the other problems on August 31, 1907 the famous Anglo-Russian Convection had been signed between England and Russia in St. Petersburg and on September 4th it became public at Tehran. The people didn't trust the wording or content of the agreement and this left a painful impression on the people of Persia.

The disord in Iran continued to worsen, and the Shah for the fourth time visited The Majlis and took an oath on the Quran to be faithful to the Constitution. In December, however, Muhammad Ali Shah decided to destroy the Majlis, and he ordered the Cossack Brigades commanded by Russian officers which also consisted of some private soldiers under direct command of the Shah himself to disband the Majlis, but the army had fallen into such disrepute that no one took any attention of it.

The members of parliament resumed their work and all the ranks of the people gathered around The Majlis so the Cossack army was unable to do anything, and did not dare to come near the area.

In the latter days of January, 1908 someone threw a bomb on the car of the Shah. This failed to kill the shaw, because he was riding in his carriage behind the Royal automobile. The bomb exploded on top of the car slightly injuring the French chauffer of the Shah named Varnet. The Shah suspected that this was the work of Constitutionalists, which strained his relations with the Majlis even more.

In June 1908, Russian and British Legations openly contributed to the overthrow of the Majlis and bombardment of Baharestan by Persian Cossack Brigades. Here the British and Russian Ministers called the members of Majlis and threatened to intervene if the Shah's plan and wishes did not ease. In early June the Shah changed his residence to Bagh-e Shah and invited some leaders of the Constitutionalists to come to him for further discussions.

The Shah then arrested some members of constitutionalists but one of them escaped and reported to the Majlis. This affair caused disorder in Tehran. The Shah continued his threats against the Constitutionalists, and he put colonel ***Liakhoff*** in supreme command of his army at Bagh-e Shah.

He then sent some Cossacks with an ultimatum to the Majlis, threatening the bombardment of Mosques if the people in the premises did not disperse. On 23

June the Cossack troops set out to occupy the Majlis. They stationed guards in the streets adjacent to the Majlis, and the troops opened fire on the Majlis buildings, killing a few nationalists who were in the premises.

The other armed Nationalists present in the Majlis opened fire destroying three of the Cossack field guns. Cossack reinforcements arrived, and despite the small number of Nationalist defenders of the Majlis, they kept up a stout resistance for about eight hours. After the Majlis was badly damaged, many of the Nationalists were arrested and imprisoned, the Russian troops looted the houses of the people that they thought to be Nationalist.

Meanwhile riots had started in provinces such as Rasht, Kerman, Esfahan, and Tabriz, from which all demanded deposition of the Shah, and a force of 300 cavalry was dispatched to defend the Constitutionalists in Tehran.

In Tabriz which was the next important city after Tehran, street fighting broke out between Nationalists and Constitutionalists, and a representative of the inhabitants of Tabriz announced opposition to Mohammad Ali Shah, and they expelled the Royalists from Tabriz.

Bakhtyari Involvement in the Constitutional Crisis

On January 1909, two chiefs of the Bakhtyari tribesmen, Samsam al-Saltaneh and Zargham al-Saltaneh, with about 1200 cavalry troops were in possession of the city of Esfahan. They had successfully ejected the Royal troops from Esfahan. The Bakhtyari chiefs decided to take the side of Nationalists. The situation got worse during the month of January and many Tabriz inhabitants were dying from hunger or barely subsisting by eating grass. The city of Tabriz was surrounded by the savage tribesmen of Rahim Khan who were aiding the Shah's troops. The Shah had promised them the opportunity to loot the city after they succeeded in conquering the city's defenses.

Two foreigners joined the Nationalists. Their names were Mr. W. A. Moore, an English journalist who was sent to Persia by several English newspapers, and Mr. H. C. Baskerville, an American Presbyterian Missionary teacher in the boy's school in Tabriz. Unfortunately on 21 April Baskerville was killed. When the situation became worse the Shah's commander of the troops sent a message to Tabriz and offered safe passage for all foreigners who would vacate the city. Most of the foreigners refused to go.

On 20th April, Russians decided to send troops in to the city of Tabriz to safeguard the entry of provisions, to protect the foreign consulates and their subjects including their own Russian Consulate in Tabriz. On the 29th April, Russian forces of four squadrons of Cossacks and three infantry battalions with batteries of artillery entered the city. They gave assurances that the troops would remain only as long as might be necessary to guarantee the security of the lives and property of foreigners and consulates and their subjects and also that they would abstain from taking part in the political conflict.

More serious difficulties occurred with Russian troops in Tabriz and the inhabitants requested withdrawal of their forces. In the meantime, Nationalists occupied a section of the road leading from Caspian to Qazvin, and they blocked the road onwards to Tehran. The progress was slow and they were waiting the concurrent advance of the Bakhtyari forces from Esfahan and the south.

With the lack of communication available in 1909, everything appeared to be in total disarray and confusion.

The Shah, acting on advice of the English and Russians, again promised to restore the Constitution, but by now the leaders of Nationalists and people had lost confidence in the Shah's pledge. The Bakhtyari troops, commanded by their chieftain, the Samsam al-Saltaneh and Sardar Asad, advanced toward Tehran. They had joined the Nationalist forces on the road to Tehran. Qazvin fell into the hands of Rasht Nationalists. Throughout, the moving spirit of this expedition is said to have been an Armenian by the name of Ephraim Khan.

On June 16, the Bakhtyari forces, composed of 800 cavalries resumed their march to Tehran. Shortly afterwards they came in contact with the Nationalists, at Qazvin. The British and Russian Legations tried to deter the Bakhtyari forces from their advance to Tehran by negotiation, but were without success. On June 23 the Bakhtyari forces reached Qum, about 80 miles from Tehran. In spite of repeated threats from British and Russian Legations, Sardar Asad replied that he had certain demands to make on the Shah, he continued his advance and refused to stop.

The Shah's troops at this time reported to be about 5000 at Saltanatabad and 1350 Cossack Brigades. The Russian Colonel ***Liakhoff*** with 800 troops was waiting for the Bakhtyaris to approach. On July 3, the forces of Karaj fell back to Shahabad, and next day a skirmish between these forces and advancing Nationalists took place. The Persian Cossacks under command of Captain ***Zalpolski***, and two other Russian officers posessing three field guns, closed with the Nationalists, and the Cossacks lost three men and an officer.

Russia sent about 2000 reinforcement troops from Baku on the 8th July and they reached Qazvin on 11th July. On this day an engagement took place between Bakhtyari troops and a Cossack Brigade at Badomak village, but the result was indecisive. Skirmishing continued for the next two days and on July 13 two Nationalist forces to the utter surprise of the Cossack Brigade and Royalist troops, slipped through their lines and quietly entered Tehran. The skill of the manoeuver is said to have undoubtedly been due to Ephraim Khan.

There was street fighting in Tehran during the entire day, and the population of Tehran received the Nationalists forces with greatest enthusiasm. While the Cossack Brigade under Colonel ***Liakhoff*** were still besieged in their barracks in the center of the city, he wrote a letter to Sepahdar Azam proposing safe terms for surrender of the Brigade. On July 15 they were in full possession of the capital, but the Cossacks troop still held the center of the city.

On 16th July, the the Shah took refuge in the Russian Legation by previously obtained consent of the Legation to his decision. Both Russian and British flags were hoisted over the Russian Minister's home as soon as it was occupied by Shah. In the meantime Colonel ***Liakhoff*** completely surrendered to the Nationalist leaders and accepted formally the ultimatum to serve under the new Government, and work under the direct orders of the Minister of War.

Late the same evening an important meeting took place in Baharestan palace, and by formal ceremony the Shah was deposed. His twelve year old son Sultan Ahmad Mirza was proclaimed his successor. Finally, the apparently lost cause of constitutionalism in Persia had been suddenly revived and restored.

A new cabinet was formed with Sepahdar Azam as the Prime Minister and he was also appointed as War Minister. Sadar Asad Bakhtyari, was appointed as Minister of the Interior. After some time Sepahdar Azam resigned and Samsam al Saltaneh Bakhtyari replaced him as Prime Minister. After a year, due to his poor health condition, Sardar Asad resigned and travelled to Europe for treatment but was not very successful and when he returned to Tehran he was still in very poor health and eventually he passed away. When he died, he was completely blind.

Samsamu's-Saltana Head of Bakhtyari Tribesman
and Prime minister with his bodyguards (in white hats)

Sipahdar-e-Azam and Sardar-e-Asad
Captured Tehran in summer 1909

Sardar-e-Asad Led Bakhtyari tribes in capturing Tehran from Muhammad Ali and Cossack Brigade in 1909

Sipahdar-e-Azam was suspected of Conspiring with the Russians before their overthrow at the hands of the Bahktyaris' in 1909

Sipahdar-e-Azam and staff

Sardar-e-Bahadur (son of Sardar-e-Asad)

Sardar-e-Bahadar with Ephraim Khan Leading Bakhtyaris against the Shahsevens

Nasiru'l-Mulk "Regent of Persia" with Bakhtyaris
in front of Parliament (February 1911)
Included are; Samsamu's-Saltana, Sipahdar-e-Azam,
Muashamu's-Saltana, Amir Azam, Sardar-e-Asad and Sardar-e-Jang

Amir-e-Mufakhkham

Amir Mujahid

Sardar-e-Zaffar

Sardar-e-Jang

These Four Bakhtyari Chiefs or "Khans" were instrumental in overthrowing the "Majlis" in December 1911

*Chapter note: All Bakhtyari photographs used are in the public domain, being in possession of the Bakhtyari museum in Tehran, or published before 1910, and the copyright lapsed

Chapter 6

The Unforgettable Meeting

It was the month of May, the summer had already begun and air-conditioning was working day and night. The temperature in the month of May is as high as 38 to 39 °C (100 degrees F) and usually rises from June to August to over 50 °C (122 degrees F).

I was deep in my work preparing the end of the month balance sheet when my secretary knocked on the door, came in and said “there is a tall American asking for you”. I asked her to show him in but the tall gentleman was already half way into my office having heard me and said “Good afternoon, are you Mr. Masoud Akhbari?” “Yes” I replied, “What can I do for you?” I got up to take his hand stretched towards me across the desk, looking at him with astonishment; the visitor was a handsome, tall, erect and powerful built man, with a narrow hip and broad shoulders, a very impressive figure; his tough and confident look made me like the man from the first moment.

I took his hand and said to him "Please take a chair". He began to introduce himself.
"My name is Blaine Taylor, and I talked to Mickey Meroney few days ago and asked him if he knows someone to help me get a hunting licence and eventually my weapons from the U.S.to here". He continued, "I am a keen hunter and I would like to hunt in Iran, Mickey Meroney said if there is any one to help you it would be Masoud; that is why I am here to ask you for a favor”.

His sharp looking eyes and frankness were deeply registered in my heart and I said to him, "When do you intend to bring your guns?” He replied "I worked for Schlumberger and now my assignment is over and the Oil Company has engaged me, I will be going for two months on vacation then in October I shall return to Ahwaz to start my new assignment with OSCO (Oil Service Company of Iran)".

The Oil Company headquarters in Iran was first situated in M.I.S. and in 1970 was transferred to Ahwaz, a city with about 3 to 400,000 inhabitants; it is situated in South West Iran. Ahwaz is about 120 km from Abadan and 130 km from M.I.S. The city of Ahwaz is on the plains of Khuzestan. Iran has a total area 1.648 million sq. km (comparative area is slightly larger than Alaska). Iran’s borders with Azerbaijan on the north are 435 km, Armenia on the north-west, 35 km, Iraq on the west, 1,458 km, Afghanistan on the east, 936 km,

Turkey on the north-west, 499 km, Turkmenistan on the north-east, 992 km and north west Azerbaijan at the far north, 179 km. The total borders with other countries are 5440 km, not counting the Persian Gulf on the south and Caspian Sea on the north comprising about 2440 km.

Agriculture accounts for only 20 % of GDP, and the principal products are rice, wheat, barley, sugar beets, cane sugar, fruits, nuts, cotton, dairy products, wool and caviar.
The Shatt al Arab is a navigable waterway about 950 km long stretching from the Persian Gulf north along the Iran-Iraq border. Larger vessels can enter only up to 130 km, of this waterway due to the depth. The most used ports of Iran in Persian Gulf are: Bandar Khorramshahr, Abadan (mostly destroyed in Iraq/Iran war 1980-1988) Bandar Abbas, Bandar Bushehr, Bandar Imam Khomeini (former Bandar Mah-Shahr), and on Caspian Sea: Bandar Turcoman, Bandar Anzali, Rasht, Ramsar and Chalous. You have probably realized that "Bandar" is the Persian word for port.

Iranian terrain is generally rugged and mountainous, from the highest altitude of approx. 5600 meters in the Elburz mountains, Alvand about 4800 meters and the Zagros about 3800 meters with a central basin and huge deserts and limited interrupted plains along the Persian Gulf and Caspian Sea. Only about 11% of the area is forest and wooded lands. Arable land makes up about 8%, comprised of about ¼ meadows and pasture lands, and ¾ irrigated farmlands. The Great Kavir (desert) covers the east-central part of the country's the climate is mostly arid or semiarid, and along the Caspian Sea coasts subtropical.

Blaine was Geophysical Engineer, his first assignment was with Schlumberger in Ahwaz South West Iran, and after two successful years or at the end of his term he was engaged directly by Oil Company (OSCO), while in Iran he showed lots of interest in the culture of Iranians and wanted to know Iran and Iranians better, he started to learn the Persian Language from the first moment he put foot in the country and was welcomed by most Persians who really happened to know him. Some talked with lots of interest about Blaine at his work and many of his expatriate colleagues loved him and enjoyed working with him. From the very first days of his arrival in Iran he developed an interest in the nation's psychological and cultural aspects and during his entire stay in Iran he was learning the language sufficiently to find his way around.

I said to him "you can bring your weapons with you. Upon arrival at Mehrabad Airport in Tehran, declare the weapons, leave them with the customs officials and obtain the receipt. Bring it with you and give it to me when you are back in Ahwaz." I could see from his expression on his face that he could not believe me! With astonishment, he asked, "Is it true what you say that it will be sufficient for me to import my guns from America so simply?" "No, that is not enough, you must provide me with your particulars and home address and four passport size photos, and you have to fill in a form which I will get for you on my next trip to Tehran, from the officials, during your absence in America. Before he left he said "Masoud I am really happy to know you, one last request I have, would you take me hunting with you if you have time? It was strange, almost telepathy, that at the same time I had in mind to ask him if he would be interested to go hunting with me when his weapons were released from the customs.

With his appreciative eyes looking straight into mine waiting for my reply, I must admit that the first impression of this man was deeply registered in my heart and I do not know what had happened, but I felt I had known this man for all my life time, I said "Of course by all means I would love to have a good friend and a good hunter with me that I can learn from him like a brother". I could see that with lots of happy feeling he left my office and next day he brought all the papers and details needed for importation of his weapons. During his absence on leave I prepared all necessary formulation and presented it to the department concerned in Tehran on one of my trips. Blaine arrived in Ahwaz after his vacation and he did not waste any time; he was in my office the day after to hand over the customs receipt and gun papers. It did not take more than a month when all necessary formalities were completed to release his weapons from the customs in Tehran, the only thing left was for him to go up to Tehran Customs in person and get his guns.

There are many species of wild game that you can hunt in Iran, such as Uryal Sheep, which are wild sheep with rather thin based Antlers which can get up to 40 inches.(see picture) Wild Boar lived in the bush, especially on the banks of the Karun and Dez rivers. Karun is a river about 350 miles long and in some places the width of the river is as broad as 1000 yards.
The very first navigation on these waters dated to early 1900 when people began to use boats to carry heavy loads.

The Stern Paddle ship was first used by Lynch Brothers, one of the pioneers, and a contractor for the Anglo Persian Oil Co. in early 1908. With it they succeeded in opening up the Karun waterways for commercial transport and to ship heavy equipment for the oilfields from Khorramshahr to Ahwaz and from Ahwaz to Dare-I- Khazineh (D.I.K.), carrying cargo which was not possible to transport by animals. The latter destination served the needs of the D'Arcy Oil Co. later called the Anglo-Persian Oil Company.

Oil Company Stern Paddle ship in D-i-K (My father leaning on the post with hat on)

Iran, some say, is a "*paradise for hunters*": Mountain goats (Ibex), Partridge and Black Partridge, Sand Grouse, Antelope, Brown Bear, Yellow Deer, and many other species, including the wild Boar in Khuzestan. Uryal Sheep are believed to be the largest types which live in Iran *(similar to Argali Sheep in Mongolia and Marco Polo Sheep in the Hindu Kush Mountains, they get as heavy as 450 pounds).*

The wild Boar feed on vegetation and cultivated fruit growing on both banks of the Dez and Karun rivers. There are Black Partridge, Sand Grouse and Quails in plenty which you see in every corner of the road just outside the cities and in most parts of the Khuzestan area.
Ibex and mountain goats are also there but to hunt them you have to travel longer distances and climb to the higher altitudes, rocky cliffs and dangerous terrain, difficult to come close to them, even if you are in good physical shape, If you can allow for wind directions you may get as close as 200 yards from them.

There are many Antelope with beautiful antlers. They also have lovely brown eyes which was one reason we never found it a pleasure to shoot them. Brown Bear lived above 8000 feet altitude and sometimes you see Black Bear amongst them but most of the time you find Brown Bears only at higher altitudes in Zagros Mountains.

Wolves and Hyenas are rather in excess and bother the villagers and the inhabitants of the black tents, (migrating people) very much. Now and then they attack a village and kill their sheep and chickens, mostly at night. Sometimes even young and old people are killed by the wolves in the winter. The village people begged us to relieve them from the threat of these animals whenever we were in the vicinity of their settlements.

Chapter 7

First Time With Bakhtyaris

I remember the first time we took a trip to Qleh Tul and ***Izeh*** for a long weekend hunt we arrived at a village at sunset in one of the villages of ***Izeh***. We took a rest and were invited by the head of the village for dinner. This man, for our honor immediately ordered the killing of a small lamb but before it was butchered they brought the lamb for us to see. Blaine was against it and he insisted that they should not kill the lamb, but I did not translate exactly what he said to the Khan, I just explained that Mr. Taylor says we have enough food with us and we do not want to impose a burden on you. Should I have translated that Mr. Taylor is unhappy or does not want to accept that you are going to kill a sheep it would be very disappointing to the Khan and they would not appreciate it very much as he was doing this in our honor and all his subordinates respect what he is doing, therefore he would not like to lose face.

Qaleh Tul, near Izeh

After dinner was served a man asked us if we would help his villagers kill some of the wolves that come to their village every night and kill or wound their sheep and goats and chickens, and make litter everywhere. They were afraid of them because they might also kill small children if they got the chance.

During that night we stayed in three different spots on the outskirts of the village facing the hills with 150 yards between each other waiting with our

spotlights ready. It was a very dark night, we could see nothing without our spotlights, and for the first part of the evening we had seen nothing moving. Just after midnight we heard someone calling from the far side of the village that the wolves are coming. At first I was very disappointed that with the man calling he was making so much noise I was afraid that he would scare them away, but the head of the village explained to us that these wolves were not afraid of men shouting. The wolves are sure that they have no enemies there so that when they come, in spite of many people going out and making noises they do not change their place of entry and would still come into the village.
I could not tell exactly how many of them there were, but only 50 yards away we saw four wolves approaching the village with grace, looking around sniffing the air and walking without hurry. The leader had white fur, and all of us were holding fire waiting for Blaine to shoot first as was agreed.
The first shot brought the white wolf down and the others also shot two more. The fourth one got away wounded but was found dead less than a mile away next day.

The dead wolves were then cut into pieces and each family took a part of the body, teeth, claws or eye etc. to hang on the baby's cradle to keep away evil spirits or as Persians say the ***evil eye.***
Some took the gallbladder of the beast to dry and hang on the cradle believing this will give the baby courage and will be brave when grown up. It was late and we needed some sleep before we could get ready to hunt in the early morning. The next day we left the village two hours before daylight, heading for the far mountains.

Chapter 8

Short Hunting Trips

The office hours of the Oil Co. in summer time were from 7 a.m. to 13:30 p.m. and after that people used to go home for lunch and rest due to the excessive heat outside. Hunting started on a very hot day of August the 15th, so I called Blaine in his office and asked him if he would like to go for a half day trip hunting wild sheep in the nearby Mushtaq hills close to the Iraqi border which was barely more than 2 hours' drive from Ahwaz.

Blaine was an honored guest for me; I could tell that he was happy to go on short trips, so we took off right after office hours, always having our entire gear ready in the car from the day before. There was no need to go home and change; we changed when we were in the hills. With one and half hours' drive, taking lots of water with us, on the way we called upon an Arab village for the boy Mazi, who was an Arab guy we used to carry our water bags. The only thing of concern in such heat is that you need to drink a liter of water every half hour of walking in the 50 degree centigrade heat and dry weather of Khuzestan.

We ended up that day shooting two wild sheep under the baking sun without wasting any shots; Blaine got his at the range of 150 yards, he could tell me exactly where his bullet was going to hit before we reached the dead animal, and always he surprised me with his accuracy.

Dr. Harry Fathi, Orthopedic Surgeon at Ahwaz Hospital bought himself a rifle recently released from customs and asked if it would be possible to join us hunting wild sheep.

As a matter of fact we did not mind having another friend to accompany our team.

On our next trip to Mushtaq hills [1] Harry joined us. We started as usual at 13:30 p.m. and we got to the hills at 15:30 hours. After walking for about ten minutes I spotted a ram with a good size horn which was about 600 yards away from where we stood, I showed it to Blaine and said there must be more game around there. I pointed out how we should go towards the animal, and while we were walking and talking on every bend we checked the area to look for other sheep and making sure that no other animals could see us.

We had not gone half the way when we noticed that Dr. Fathi was missing from behind us. We did not know whether he had remained where we spotted the sheep or not, and due to the hot weather conditions and the existence of dangerous wild animals we decided to go back to where we started to be sure nothing had happened to Harry.

We were about to reach the spot where we left him when we heard a rifle shot, so we ran over the hill and we saw that Dr. Fathi still had his rifle in his hands pointing the muzzle exactly towards us! Blaine gave a shout, and when Harry saw us he called out," I've shot one, I've shot the ram, get ready, the others are running towards you", but the game had already seen us, they changed direction and began running away from us in a straight line about 500 yards towards the far hills.

Blaine aimed and shot, and while I was watching them with my binoculars I saw one that stumbled, got up and then suddenly disappeared behind a hump. I thought that Blaine had missed, asked him if he was sure he shot the animal. Blaine said "yes, absolutely sure, I could even see where my bullet hit, in the chest. One should always have control over both eyes and be able to aim with the right eye and see at the same time with your left eye, so that the movement of the rifle doesn't cause you to lose sight of the world around you."

In Iran we learned to shoot with the left eye closed while the right eye was used for sighting, and this caused many problems.

While Blaine was looking through his binoculars he suddenly froze and very quietly said, "Masoud, I want you to look at this one, a beautiful Leopard with a long tail walking softly and sniffing the ground on the same trail over which the game had gone. Blaine said we have to hurry to get there before our game was gone with the hungry Leopard following them... Unfortunately when we started to walk towards the animal Dr. Fathi came out from the opposite ridge and calling at the top of his voice that we have missed our aim, not knowing that Blaine had shot his game already.

As a boy I used to walk with my father and carry his back pack while he was hunting with his long muzzle loader, I was very familiar with these hills and I knew every corner of them.

I remember one day very early we set foot to go hunting; my father was walking before me when he spotted few wild sheep. He asked me to go over to the opposite plateau and sit there until he had taken a shot, then I should return to him.

The place where I sat was on a higher spot and with my binoculars I could see my father walking towards the game. While I was lying there stretched out I felt something sniffing at my legs, suddenly with a swift move I jumped up to face a hyena which was shocked by my sudden movement and jumped back a few yards staring at me and then began walking away. Thank God that he was not hungry that day.

There was one thing in particular I admired about Blaine; he never got angry or nervous at all and he was always calm and accurate, watching his steps. Blaine was a sure and accurate shooter and he knew everything about weapons and firearms and hunting. He was a great help to me and a good teacher on how to handle my weapons. He had a particular passion for making sure that everything will go according to his plans and that nothing would go wrong while we were in Iran.

Blaine was feeling at home with the Iranians and I could feel that he, although his culture and habits differed from theirs, still felt safe and confident among those proud people.

He also succeeded to win the heart and friendship of many Iranians especially the Bakhtyaris, with such facilities that made him famous among them, he was respected and they talked about him to others and while travelling they would come a mile or two to greet Blaine and welcome him to their community of black tents which were all over the mountain area.

Blaine always carried some sort of presents to give away to the poor or his hosts while on such an expedition. He was brought up in Malad, Idaho, U.S.A., in a simple and respectable family, his father was also a great hunter and most of the winter he was trapping fur and in summer hunting for meat. Blaine unlike most others was fond of the outdoor life, he believed in freedom and expressing his thoughts and views quite openly. On most trips Blaine accompanied his father from the age of ten with his rifle and shotgun and used to participate in bringing home game for food.

Chapter 9

Blaine And Masoud

The ties between us grew tighter and closer every day and we started to know each other very well as time passed; sometimes I would feel that Blaine could read my mind or I could guess what was bothering him when he was deep in his thoughts. I recall exactly my feelings on the first day we met as well as if it were yesterday. I was sure even then that we were going to be great friends, and as Blaine later quoted, "brothers" in a pre-existent or spirit world, and had been separated by distance in this world, but yet distance did not stop us from meeting each other regularly and hunting together in different parts of the world.

Blaine, I discovered, is one of those calm, relaxed men who are capable of tremendous outbursts of dynamic action such as I know in his type well. It is always our custom to respect a request even if we don't know the person, but we managed somehow to develop our friendship in these lonely valleys and high mountains of Zagros where I came to understand for the first time and to appreciate the real power of friendship. How much friendship has power and value to influence your soul, one can't estimate, but the quality of true friendship is never appreciated unless it is lost. True friendship will absorb you with such power that you never want to separate yourself from it.

Blaine was saying that I had travelled over half the world to find my half-brother, regardless of the religious difference; Blaine was a Mormon and I was born into a Muslim family, but our opinions and understanding were indeed alike and our feelings were close like two brothers. I have come to the conclusion that what Blaine quoted sometimes during a lecture or predicted amongst friends and people that we knew was absolutely correct. Later incidents proved that there is more substance to this than we know of.

By pure coincidence, my eldest son at the age of 18 finished his high school in Iran and told me he wished to go and study at a university in the United States. The very day before I decided to ask Blaine for his advice, my friend called me and said "I shall be going to America in two days' time and I had in mind to investigate your son's entry to Utah State University!" What a coincidence, he must have heard us talking about his study in the USA and understood my thoughts so he taken the first step. He said, "Your son can fly to Salt Lake City while I am there and I can arrange his entry into the university". This was done as planned and Blaine helped my eldest son enroll in Utah State University where his mother was a senior university librarian. She was a lovely lady who was so kind and helpful to my son, she really took good care of him and used to invite him regularly to her house in Logan.

Such a coincidence again happened during later years; my son at the age of 28 when in Salt Lake City at a job sponsored by my friend (brother) had sudden fatal cardiac arrest while having a drink in a restaurant. They called me late in the evening and my wife called Blaine who happened to be working in London. What he said in reply was "not to worry, I will go to Salt Lake City tomorrow morning and arrange the funeral for him".

What can you call this, a friend or a brother? " If I say it means to me and my family more than a brother I won't be wrong. He was as he said next day in Salt Lake City and he made all the arrangements for my son's funeral in a very short time.

The quality of my shooting and hunting improved due to Blaine's guidance and instruction on how to use my weapons accurately and was so good that even if a mountain sheep were running with a speed of 40 to 50 miles per hour I could shoot it down accurately.

We were asked by many Oil Company officials to kindly accompany them for hunting.

We did take many short trips away and we came back with lots of game. During one of these trips we had to rest at mid-day in a cool cave. When Mahmeed came later and put his rifle down Blaine asked him to empty the rifle before sitting. Mahmeed said "the rifle is empty". To prove this to us he got hold of the rifle and pulled the trigger. The whole world seemed to have dropped onto our heads, the cave was completely filled with dust and lots of gravel and tiny stones were flying and dropping from the ceiling. "What a fortunate accident," Blaine said, "good thing Mahmeed didn't have his muzzle pointed towards us otherwise one of us would be dead by now!" This proved that Mahmeed had no proper upbringing with guns and safely handling of weapons and as Blaine said it was dangerous to let him use weapons before he learns how to use them. This incident made Mahmeed decide to learn from Blaine on every short trip they made all about gun safety.

Such incidents always happened when we took off in the hills or mountains with people that did not have proper instruction or education on how to handle their weapons but so far we were lucky to escape from any serious injuries. One other danger was snakes and scorpions, there were so many to be seen. Some were so poisonous that they would kill a person in 3 minutes if they bite. That means you will have no chance to reach hospital after you have been bitten by them. During his stay in Iran, Blaine and I made many short trips and weekend hunting expeditions which made us closer companions each time.

One night we decided to get some wild boar, we drove about 10 miles away from city of Ahwaz close by the sugar factory. It was midnight when we got to the place and our spot lights lit the area near the river banks while driving up and down on the edge of cultivation, but no wild boar were to be seen. We drove around until about four in the morning but we saw not one. While driving, Blaine ran over a Peyton snake which is very dangerous so he backed up and put the headlights on it. The snake was trying to crawl away, but with a broken back it could not do much, so Blaine got hold of a wooden stick and started to move the snake and raise his head. Suddenly the snake bent over backwards and stung its spine it was dead in less than one second and stayed motionless.

By this time we were really tired and I suggested to Blaine to stop and have a nap before we went home. I don't know how long I had dozed off when I jumped up with the sound of a shot in my ears, my head hit the side of the door and I did not know where I was, I asked in a loud voice what had happened, but Blaine did not talk and started reloading his gun and aiming, by this time I was wide awake and I looked to the direction Blaine was shooting, I saw the whole field in front of us was moving and with astonishment I was still busy looking at them when Blaine shouted at me saying "what is the matter with you, can't you see those wild boar running wall to wall? Time to wake up and get some.

One of the 650 pound Wild Boar Blaine shot

Another large Wild Boar Blaine shot

What we did normally, after having shot more than the required number of wild boars, we took them door to door to each of the expatriate houses. Then we would ring the bell at four or five o'clock in the morning and keep pushing the doorbell in until we heard finally someone was coming to open the door all the while shouting and cursing who can be so early in the morning, then clear from there before the gate was open we would drop the animal in front of his gate and get. We did this because the expatriates, who knew we were hunting often, had asked us for a wild boar. They just didn't know they were going to have it delivered whole, freshly killed, at 4:00 o'clock in the morning. They had only pictured packages of meat neatly cut and wrapped and frozen.

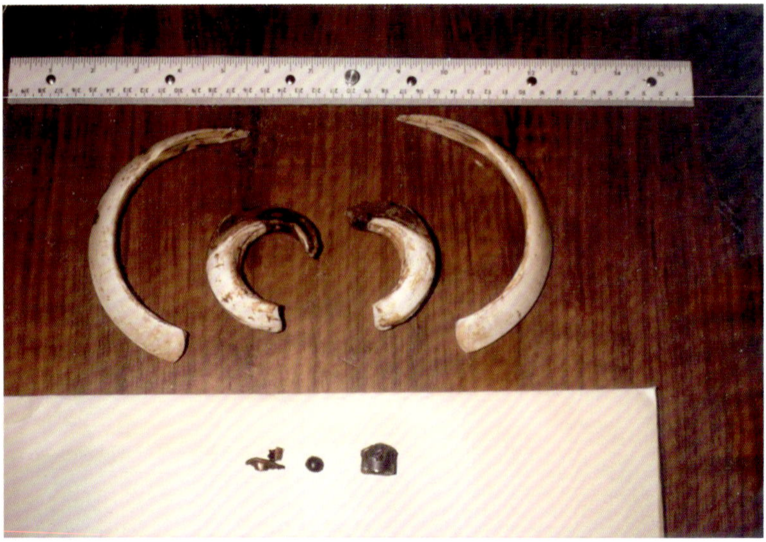

Upper and lower tusks from a big boar, along with bullets found when butchering

The next day Blaine and I received lots of telephone calls. They had figured out who was there early in the morning waking them up. Of course they did not know how to clean and prepare the boar so they wanted us to help them by skinning it for them, and showing them how to cut it up. Instead, they gathered at Blaine's home for a class in the basic skills of skinning and cutting up a game animal.

Once Blaine and Bill Whitney asked Mahmeed to go hunting in the mountains of Bibi Hakimeh, Mahmeed spotted some wild sheep. Blaine made a habit of always giving the chance of first shooting to the person that was accompanying him, and when they spotted wild sheep they were quite on top of them not more than about 50 yards away. Bill aimed and shot three rounds and nothing happened, by this time the game was about four hundred yards away running in a straight line away from them, Blaine aimed and took a long breath before he made his shot, moreover he was waiting for a good moment to shoot surely.

Bill said "I don't think you hit it so it is not necessary to search for it there, we'd better go back to the car", but when Mahmeed said that while he was watching through my binoculars he saw that Blaine had shot a sheep, they decided to go and look for it. Mahmeed found the sheep, as he said in the first place, and the animal was shot right through the neck at the back of his head.

The following funny incident is not out of place if we mention it here. Bill Magnus was at that time General Manager of I.O.E. & P.C. and used to hunt as well. He was very fond of Black Partridges, he used to give a cocktail party and dinner for most of the staff every first Wednesday of the month, and after the cocktail party was over, the senior staffs was invited afterwards to go to the dining hall, inside the number 1 bungalow, the residential house of Bill Magnus. He used to have among his various food some wild game dishes, which either he shot or we used to provide; Wild Boar, Black Partridges and Quail sometimes for such occasions.

Normally the chief cook used to mark the dishes with a label showing what kind of meat they were serving, but the very first time the cook presented the Wild Boar to the guests he did not mark the plates and some of the high ranking Iranian officials of the Oil Company were Muslim. Even though they consumed alcoholic drinks they did not eat pork on principle because in Islam pork was considered unclean and it was banned from human consumption.

After the dinner was over and tea and coffee with liqueurs served some Iranians stated that the special dish of meat was the best roast meat they had ever had. I recall this incident clearly because Bill Magnus got up and said loudly that the meat was indeed delicious and it was Wild Boar and not beef, and that it was carefully marinated by his experienced cook. He had not finished his talk when some of the guests especially the ladies got up and ran for the rest room and pretended to be sick. Later we heard that indeed some of the ladies had thrown up.

During the time we were hunting together, I had almost burned out my 270 Winchester rifle barrel and it was not performing accurately any longer. To get a new barrel would mean going through extensive bureaucracy and formalities which would take at least a year or two if you are lucky enough to obtain your permit to import a new barrel for a gun.

One day Blaine said that he had got a friend in the American Military Advisory Department in Tehran who could obtain a faster delivery. He talked to Clyde Martinson in Tehran and indeed he arranged in no time to get a new barrel for my 270 Winchester rifle. When the barrel arrived in Tehran, one day Clyde called from Tehran and said to Blaine "the iron is here, you can pick it up". Clyde later on was the person who dedicated the grave to my son when he passed away in Utah in 1984.He proved to be another good friend and we are still friends after nearly four decades.

I should like to mention here another person with whom we formed a strong friendship in Iran ho also was interested in the Iranian culture and history; Gill

Jurenka and his wife Shirley who lived in M-I-S.; he worked for IOE &PC as Petroleum Engineer. While in Iran he loved to go hunting with us and he enjoyed sharing the Bakhtyari life

From left: Abbasi, Gil Jurenka, Jeff Chandler,Masoud and Parvizi

Gil Jurenka in the middle in Bakhtiyari clothes

Gil Jurenka between Abbasi, Parvizi and Masoud sitting

From right: Masoud, Gil, Ted Shetzer, Beckman and Abbasi

From left to right; Parvizi, Jug Bel, Gil Jurenka, and Abassi

Left to right; Masoud, Parvizi, Abassi and Gil Jurenka

A farwell party given by Masoud at the M-I-S. Cricket Club for Gill Jurenka serving two wild Sheep

Chapter 10

Zagros Steppe

The Zagros Mountains are a very important feature in Iran and in the world's geography. This chain of mountains is situated on the plateau of Iran, and goes from an altitude of almost sea level to about 4500 meters (14,500 ft.) in altitude. There are two large chains of mountains; the Alborz chain of mountains in the North and the Zagros chain of mountains stretching from north-west to south-east of the country. The Zagros range encompasses an area of 4.120.000 sq. km. Most of the heights and elevated grounds of the Zagros are covered by snow in winter and the higher altitudes are covered the whole year round. The plains of Khuzestan, Fars and Esfahan are vast fertile plains, and they are irrigated by rivers which find their sources in the Zagros Mountains. These areas are considered to be the agricultural strongholds of Iran.

A photo of Zagros Mountains, showing three levels of Mountains

Kuh-e-Safid, in the Zagros Mountain range in spring time

Streams of water flow down to green plains in different directions, and build up famous and well known rivers such as the Karun, Karkheh, Dez, Zayendeh Rud and Seymereh Rivers.

The rainfall of this region ranks second after the northern provinces of Mazandaran and Caspian area, and averages 430 mm per annum for the past 20 years.
Water sources from the Zagros provide essential support for the agriculture of the country on both sides; the higher elevations of Zagros are suitable for dry farming, but up to now agriculture on the higher parts of Zagros has not been successful.

The Zagros has a culture of honey bees due to the existence of flowered plains and many areas of vegetation, and so the area may be made to produce tens of tons of honey in areas such as Bakhtyari, Chahar Mahal, Kurdistan and the Khuzestan plains.

The migration of the nomad tribes of South Persia living in the Zagros Mountains happens twice each year. In winter, the migration to the plains of South-West Persia starts at the beginning of September, while migration to the summer area starts from the beginning of March each year. For thousands of years as long as man remembers this has been going on in the Zagros mountains; the Winter is very cold up there and with no exaggeration it reaches minus 35 C. in high altitudes and nobody can travel in the deep snow falls.

Bakhtyaris are one of most important tribes of Persia. They live a nomadic life in black tents in summer in the Zagros range of mountains about ten to thirteen thousand feet above sea level, then in Winter they migrate down to the plains of Khuzestan, Masjed-i- Sulaiman (M.I.S.) and Lali about one thousand two hundred feet above sea level.

In contemporary documents available that that Bakhtyaris are the purest Persian race. They were not influenced by other nations which invaded Persia from every corner. Due to their strategic living conditions which were almost inaccessible they maintained their purity. Non-Bakhtyaris could not travel or break through into these highlands and steppes and mountains without roads or knowledge of the climate.

When Arabs and Mongols in turn attacked Persia and conquered this land they could not penetrate into the Bakhtyari highlands and Zagros steppes. One should not mistake between Lur and Bakhtyari, Lurs are living in Lurestan which is situated in the North West and that part of the Zagros ranges.

The Lurs ambition is hunting and raising goats, some sheep and some cattle. It is very difficult to become friends with them, they do not obey any authority, and travelling in their land is not without danger. Bakhtyaris are always armed. The last Shah of Iran and his father for years tried to disarm them but were not completely successful due to their yearly migrations across international borders.

The Bakhtyaris live under black tents for most parts of the year. They are breeders of sheep and goats in limited numbers which they consider their fortune. The Bakhtyaris marry among themselves and are mostly related to each other, they have large families with many children.

Most of the men learn horse riding from their early days; women do not sit with men if they have visitors in their tent. Contrary to other cultures in Iran, they freely walk in and out shaking hands and bringing food and supplies whenever necessary, without covering their faces. In the Arab provinces of Iran, you never see uncovered women in public and you never meet them when you are invited for meal or to visit their homes.

Most Bakhtyaris who used to live in the Zagros do not have high degrees of education but by no means can you consider them uncivilized; their identity has been preserved due to the nature of the land and mountains and the difficulty of communication, and therefore they remained independent. Some of their children obtained university degrees and obtained overseas scholarships and received high posts in governmental and private enterprises.
Bakhtyaris are among the most powerful nomads of the various tribes of Persia. In 1909 they forced Mohammad Ali Shah of Qajar to abdicate by marching first to Esfahan and then onwards, into Tehran with a large contingent of cavalry, and captured the government houses there.

The first time Oil was discovered in Persia was at Masjed-i- Sulaiman (M.I.S.) in May 1908 the Oil concessions held by Anglo Persian Oil Company (APOC) were on Bakhtyari land. Bakhtyari Khans held shares in the APOC for a long time. After the British government acquired the ownership of APOC, they purchased the shares back from Bakhtyaris and also the Persian Government.

The Persians are not Arabs, the coastal parts of Persian Gulf are widely inhabited by Arab clans who had crossed from the Arabian side to Persian main lands and ports, and these Arabs sometimes intermarried with Persians and stayed there and maintained good relationship with Persians.

The Persian government in the 19th century and beginning of 20th century was rarely in a strategic position to exert its will over these coastal rulers and inhabitants due to lack of good roads and the situation of the mountains. The coastal inhabitants and rulers mostly dealt with Europeans who held their naval powers in the Persian Gulf.

Bahrain Island and Oman had been under rule of Persians and in fact based on Persian Soil in the past, the ethnic argument for historical dominance of the Persian Gulf is complex, many of these Arab population who emigrated to Persian costs and now living there though Arab in blood, but gradually integrated with Persians, learned the language, customs, food and adopted the Shiite Muslim religion. At the same time there is a major settlement of Persians on the Arab soils in the Gulf who play major roles in the commerce of Oman, Bahrain, Sharjah, Dubai, Abu Dhabi Qatar and Kuwait.

Chapter 11

The First Big Trip

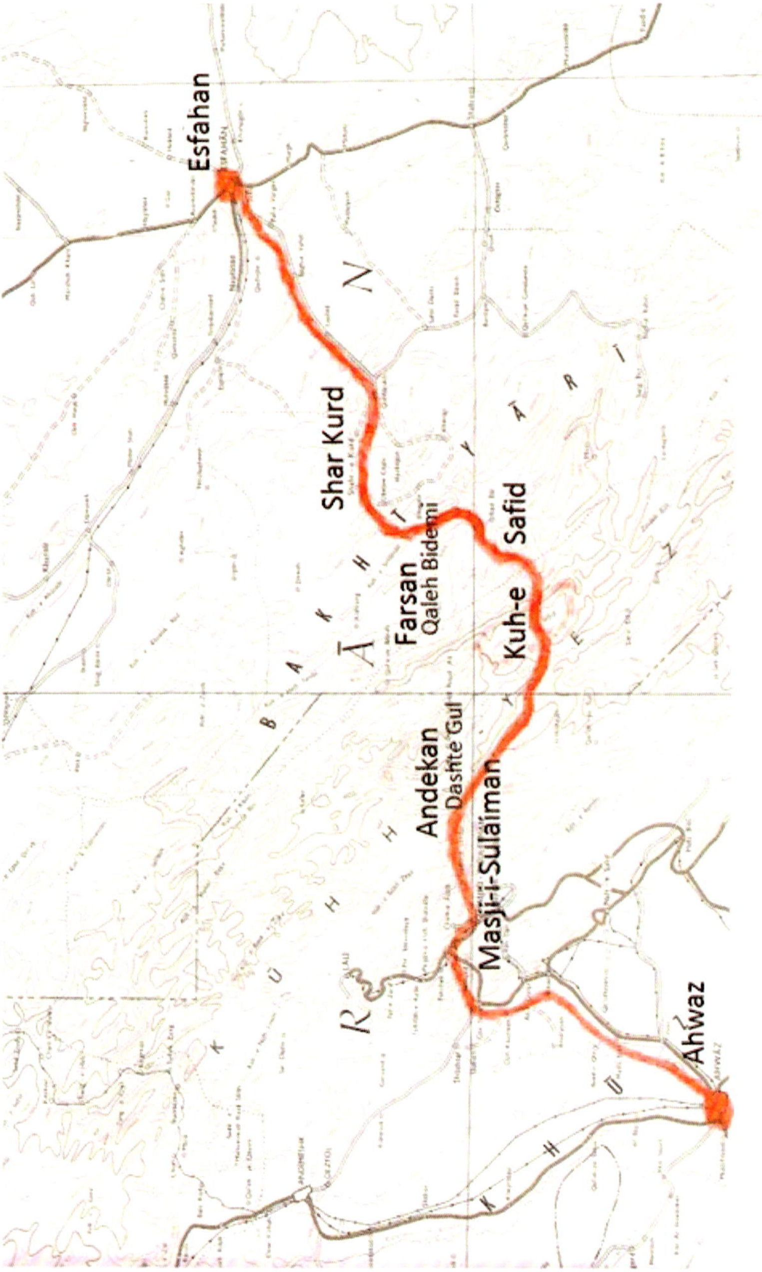

Masoud called me one afternoon in May, and said “two young Bakhtyaris came down from M-I-S inviting us to go to their place on Kuh-e Safid (White Mountain) to hunt and experience Bakhtyari hospitality”. He asked if I could take 2-3 weeks off work and join him in a trip across the Zagros from M-I-S to Esfahan on horseback with a train of mules. I immediately said I would, as I had been planning for my family to take their summer vacation in the USA. This was not Masoud’s first time to go on the Zagros trip, he had been on the Zagros many times before. The next day Dr. Harry Fathi was on the phone. He had already heard of our proposed adventure trip, and asked to join us on the Zagros expedition. He had been looking forward for many years to make this trip but he could not do it alone. We were happy to have him with us, as he was a physician, and had been hunting with us a few times.

Mr. Ali Akbar Ahooie, one of the traffic MP's of the Oil Company traffic office, took vacation from his work, to join our party as well. He was a very handy man with local inhabitants, in fact he was a well-educated Bakhtyari who lived and worked for the Oil Company. He had been on this trail many times when he was living up in the mountains in his youth.

Preparation began by securing the services of, Mahmeed, a Bakhtyari guide who had been with us for the past several years. He made all the necessary arrangements; He ordered ten mules to carry our camping gear and also he hired several Bakhtyari workers to join us for carrying the necessary fragile goods in their packs.

Judith and the children left for the USA with the annual family migration from southern Iran as soon as school was out in June. It was in mid-August when we left from Ahwaz to the house of Ahooie, who lived in M-I-S. We had a light supper and collected the last of the things which had been sent ahead of us. This was to be a one way journey as far as the mountains were concerned, and we would return by highway, once we had crossed the high Zagros Mountains. This was not an easy trip. The Zagros climbed up to over 14,000 feet in elevation, and the trip was through wilderness with no roads. The total distance we had to cover between roads, was over 300 miles.

Packing our gear on the truck at Ahooie's house in M-I-S.

Immediately after supper we left to catch the last of the daylight to drive the trucks to Godar Landar. This was the jumping off point, where the road ended and we had to cross the Karun River on floating goatskins to get to Andikan side!

I shall never forget the feelings of the group as we left M-I-S. Mrs. Ahooie threw a bowl of water on the road behind us; this was in the form of a Bakhtyari blessing to ensure our safe return. If one does not want you to come back, they would have thrown a stone into the road behind our back.

At Godar Landar the Karun River was running wide and fast, we had to stop because it was impossible to drive further than this point. Since the river here was running extremely rapidly and made a very difficult crossing we had to rely on the local Bakhtyari experts who managed to help people cross the river every day. With such rapids from the point you enter the water until you disembark, your raft on the other side of the river would gain an enormous distance down the stream. So at the bank of the Karun River we unloaded all our gear and sent the Pickup truck and truck back to M-I-S.

Crossing the river was somewhat dangerous, as the river was so swift with blocks of the snow melt still brimming the banks. We crossed with the help of the river men thereabouts, using rafts or ***Kalak***, made of pieces of log tied together on inflated goat skins set underneath. The crossing had to be done in several trips due to the large number of people and heavy gear with us. The animals swam, which they did without trouble.

Raft Crossing at Godar Landar to pick us up

Ready to cross the river at Godar Landar

On the other side of the Karun River where the Andikan area is we had to stay overnight because it had taken us longer to cross the river than anticipated. Mules were sent from Zagros to take our gear but it got dark and no one was willing to pack the waiting mules to continue our trip, because they had worked so hard to cross the river before it became completely dark.

From this point there were no roads available, just trails for the horses, mules, and foot traffic. The Bakhtyari helpers told us that the riverside was not a suitable place for camping at night, therefore we had better camp further up near the summit where there was a small settlement and flat ground for camping.

We had to put temporary packs on the mules and follow the nomad trail up the mountain which was about 2 or 3 miles. When we reached the summit we were met by two Bakhtyari men who showed us the place where we should rest for the night. After another light meal we all went to sleep in order to be ready for an early start in the morning. Even though we were sleeping not far from a very cold river, the weather was hot, and we slept on top of all our blankets, and cots, off the ground to prevent attracting poisonous snakes and scorpions.
When we woke up next morning we were greeted by the sight of such primitive housing, that we wondered what kind of people lived there.

Our sleeping quarters on the first night at Andikan

Huts were made of twigs and weeds; they wouldn't possibly keep any wind or rain out. But as always if you look deeper you will find a reason; when I questioned the villagers about this type of living conditions, they told us that there was no rain between June and October anyway, and they wanted the wind or breeze to blow through and cool them down. They splashed water on the twigs in order to cool the air. Some simply put a cloth over the hut in the daytime to keep the burning sun off, and as soon as sun was low in the sky they would uncover to cool things off. All the huts were made either of dry thorn or dry brush.

Our first breakfast on the trail at Andikan. Masoud and Dr. Fathi sitting, with Ahooie and Mahmeed standing behind.

Early morning work; Separating the wheat from the chaff

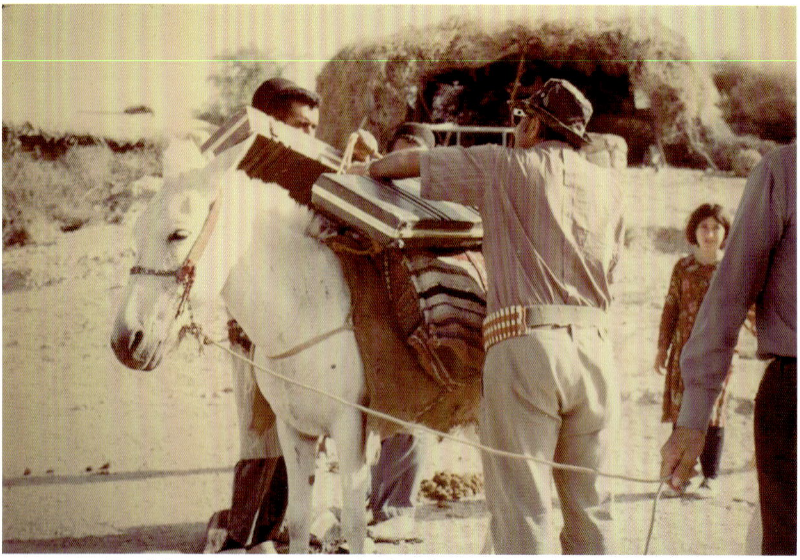
Packing our gear on the Mules early after breakfast

Many people die every year from snake and scorpion bites especially during the summer months, the snakes are extremely poisonous varieties like Cobra, Sand Krait, and Mamba. The little Sand Krait is so common, it is referred to as a" One Cigarette Snake" because that is about all the time you have left after being bitten! The first summer we arrived in Ahwaz, there was an average of almost 100 people per month who died of poison bites. This was in a city of about 300,000 inhabitants.

When we got up and around, we loaded the mules and after breakfast, it was obvious that the trip was really on, and a great feeling of excitement pervaded the camp. I was excited and I kept asking Masoud what I could do to help him, and the people we had hired to carry the freight for us. It was a real problem for me to be waited on, as I had been raised in the outdoors, and was quite capable of doing for myself. We could not keep the boys working all the time and this proved to be a constant difficulty, but in retrospect it should have been expected. These people weren't used to punching a clock; they worked when they needed to work, and rested when they saw no need of it... They knew what was in store for all of us, and they were in no real hurry to experience it again.

Walking in the mountains is never easy; if taken slowly it can be quite enjoyable, but if pushed too hard it is hell on earth. As the ***"new kid on the block"*** we expected to have some difficulties, and I did not have long to wait.

As soon as I put my leg over the mule, we had a little Rodeo display. One of the boys had put a burr under the saddle of my mule.

I was taken completely by surprise and had a difficult time staying on the mule, however, I managed to keep it under control until the saddle cinch which was not used to such rough treatment broke, and the mule and I parted company.

After being thrown off his back, the mule decided that it wanted to stomp me to death. Masoud shouted and warned me but I was more experienced with such animals than they thought, and it was my long legs that saved me from serious injury.

Masoud described it in his own words later. "I saw Blaine lying on his back. He put both feet under the mule's chest and held it off above him until the boys got it under control. Blaine received a few bruises on his legs and some more on his pride from the experience. When he got up, he was so angry that he just had to hit something or someone.

The mule was standing there with fire in its eyes still trying to get to Blaine, so it was nominated as the target of his frustration. Blaine hit that mule hard with his fist between the eyes and down it went, for a few minutes everyone thought that Blaine had killed the poor animal, and I could see that Blaine himself felt bad about what he had done, the owner of mules started to shout and turned to us for help, saying ***"Hay kor Farangi Qater mone Kosht"*** (literally means this foreigner killed my mule), but after a while the mule started to move and slowly tried to get up. Blaine said, he was displeased about such a start to the trip, and he just grabbed a new saddle and put it on the poor beast". As can be seen in this picture, each saddle had a crupper under the tail, because without it, the mule would crawl out from under a saddle in a few minutes.

Hormuz having trouble with his Mule;
look at the crupper strap under the mule's tail

The mules also had a habit of blowing themselves up with air so that you couldn't get the cinch tight. Well this was not a problem for Blaine. He was used to this, when this mule tried that old trick, Blaine was still angry, so he just kicked that mule right between the legs, and while it went "whoosh... groan" Blaine quickly tightened the cinch and put the crupper in place.

All of this must have appeared to be very funny for everyone. All except Blaine were having the time of their life. After getting the mule saddled, Blaine tried to find the cause of the mule's complaint and it was then he found the small burr under the saddle was put there by one of the boys to find out if Blaine was a good rider or not. It sure proved to them that indeed he was.

Blaine jumped on the mule and rode in a circle for about 5 minutes until the mule settled down. That mule and Blaine never did have any further disagreement, and neither did he have any more trouble with burrs under his saddle from the Bakhtyaris.

So we continued our journey up towards the high Mountains of Zagros. The road was the only trail, it has been there for thousands of years and the tribes were migrating on this trail twice a year as explained earlier, with all their family and belongings.

During their time of travelling, which took them about 8 to 10 weeks, they married, and babies were born without any hospital facilities available within 200-300 miles. They buried their dead at the side of the trail wherever convenient in the middle of nowhere, and if the man was notoriously brave amongst the Bakhtyaris they put a stone statue of a Lion on his grave to mark his being brave.

Grave yards on the ancient nomad trail near dasht-e-gol

A tomb stone of a brave Bakhtyari man

The party was now on the move, and we started to pass through ***Shimbar.***

At noon time we reached of the mountains, ***Dashte Gol*** (Area of flowers), in spring would be covered with many different kinds of wild flowers and wild tulips.

Dashte Gol (Valley of flowers)

Red Poppies (not related to white opium poppies)

Dasht-e-Gol (Valley of flowers) "upside down tulips"

The Bakhtyaris saw two rabbits running uphill at a distance of 90 to 100 yards. One of the boys turned to Blaine and asked "Mr. Tyler, (the native Bakhtyari has a difficult time saying Taylor) can you shoot the rabbit for our lunch today?" Blaine was not in the mood for shooting after a long morning ride. He was also tired from lack of sleep and was happier to stay calm in his saddle rather than shooting a small rabbit, but Mr. Ahooie trotted his mule and came to Masoud's side and said "You'd better talk to Mr. Tyler to shoot the rabbit and make a good start. If he gets it he will be respected by the Bakhtyaris for the rest of the trip and his fame will be carried all over the Zagros area."

Blaine asked me why should I shoot a rabbit which I don't want, and moreover after it has been shot nothing will be left of it I asked Blaine just do it quickly and I will explain to you later on. Blaine got his 25-06 rifle out from side of his saddle. He chose a high velocity hollow point bullet and while riding on the mule he aimed and shot.

The poor mule had never before had a rider shoot from his back and in shock bucked and threw him and his rider in the air. Blaine landed on the animal's back again sitting firm and rubbing the animal's neck to calm him down. Everyone was looking at the place far behind the target where the bullet had raised the dust, but there was no trace of the animal and the boys were shouting that Blaine missed the rabbit, but Blaine turned to them and said "when you go and get the rabbit you will see that the bullet hit exactly on his head."

The boys ran to the place reluctantly where the rabbit was shot and after some silence suddenly one of the men cried and said "you have shot it!" ***(as if he could never believe it possible*).** To their astonishment, the whole rabbit had turned pink and red, a terrible mess no good at all for barbecue any longer. They had never seen the results of modern hand loaded cartridges before. This incident kept the Bakhtyaris busy talking for 15 minutes and arguing about the distance and the resulting devastation. They paced it off, and it was 450 paces from the place we were sitting. After that we noticed every one looking at Blaine with more respect and lots of attention. They went out of their ways to please him. Blaine was the best marksman these people had ever seen, and they respected those skills more than anything else.

Shimbar is a green land and has beautiful scenery especially in spring seasons when the wild flowers and tulips are growing everywhere. It is the summer quarters of the Bakhtyari tribes when they migrate to their ***Qeshlaq*** quarters.

A picture of Shimbar valley at the foot of the Zagros Mountains

On the second day of our travel the weather was still warm in the day time especially between the high mountains and in deep valleys, so we stopped at about 2 p.m. in the afternoon to rest the animals and have lunch then waited for the cooler weather in the evening to travel. Even though Andikan is at a much higher altitude than M-I-S., it is still low compared to the Zagros, and is rather hot during the middle of the day. In fact we routinely camped each day until 5:00 in the evening when we would resume our travel. This place called ***"Dare Nargesi"*** Valley of Narcissus, in February and March each year this valley is clothed with a gay mantle of beautiful Narcissus and wild Tulips, but not for long. Later on upon those same scorched slopes the blazing hot summer sun leaves nothing but a low growth of thorns for about 6 to 8 miles.

Camp near Dareh Nargesi, Dr. Fathi sitting on the bed, under a spreading Oak tree,
Masoud standing, Mahmeed and others preparing food.

Here and there are steep valleys where the shadows fall early and rise late, strewn with deep water courses, here the eagles and other birds hover and pass with their mournful cries through the gorges. Beautiful though it is in this majestic loneliness, this country is not a place where men are tempted to seek an abiding dwelling. This early afternoon "siestas" are often scorned by people who are not used to an extremely hot climate. I think after having lived in Iran, I began to understand the people and their habits especially Bakhtyari tribes.

Blaine and Masoud deep in sleep in the afternoon after a hard day's ride and walk

There were mostly Oak trees but there are others bearing fruit such as Figs and different kinds of Berries and Walnuts. The boys would pick them wherever they could find them, and the water was so clear and sweet. We spent so much time riding that everyone knew that we would suffer from constipation; here we spent about a half an hour collecting nature's laxative figs and berries.

Our traveling companions after collecting figs. The two on the right are having a laugh, because of what Cheragh Ali is holding in his hand, while Hadad has collected figs in his shirt, Cheragh Ali was lazy and didn't want to climb. When we got the camera out the quick witted man pulled out a box of matches to show us he was ready to make a fire.

We continued our travel again until we were near the Holy Shrine of ***Sultan Ibrahim***, the place was situated near a green meadow with lots of water, and running springs, this was also another rest place of the nomads during their migration. During the migration season nomads stop here to pray and pay their respect to the shrine of ***Sultan Ibrahim*** and also rest for few days before continuing their journey up or down. Immediately after the helpers finished with their unloading, they led the mules to the meadows and left to visit the Shrine and bid their prayers.

A water operated flour mill near Sultan Ibrahim. Note the Water channel feeding down into the mill from uphill.

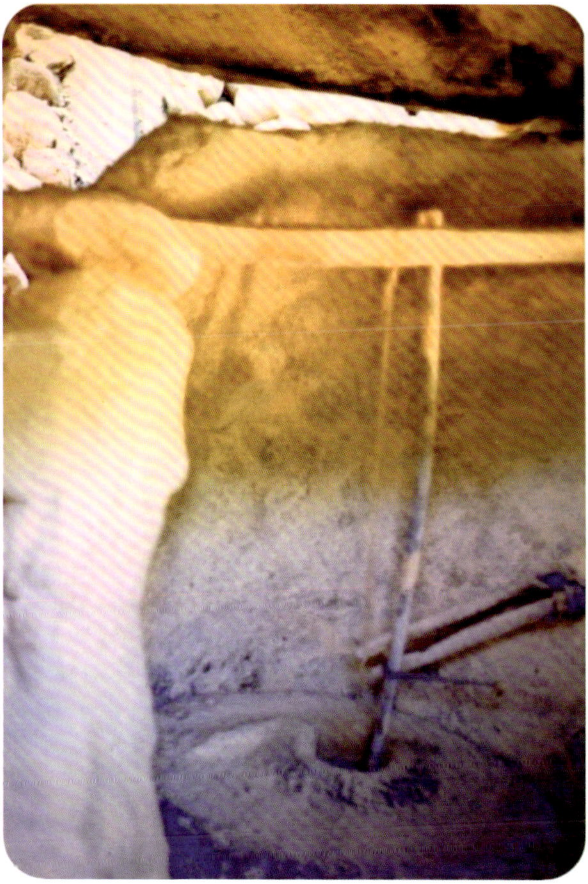

The water driven stone in the mill house

Chapter 12

On The Nomad Trails

Blaine after having lived in Iran, his familiarity with the mediaeval way of living and thinking of Bakhtyari tribes made him feel a close affinity with them and he began to understand the people and their habits especially those who lived in the mountains.

There were plenty of wild fruit trees, such as figs, different kinds of berries, walnuts etc., and the boys would get them where they could find them. The water was so clear and sweet that is was a great surprise. We continued our travels into ***Dasht Ahangeri*** (***Valley of Blacksmiths***).

This valley was a common wintering place for the nomad tribes, where they re-shod their mules and horses in preparation for the long trek the following year.

Masoud, Blaine and Dr. Fathi had camped in the ruined fortress far right, the night before

On the way near the road we passed graveyards which were unique and interesting for Blaine. When a member of a family dies on the road during migration, they bury the person in this graveyard in the middle of nowhere; the only mark they use is a face of the man or woman carved on the stone without name or any inscription. After all the family disappears from the region the remaining people would re-use the grave markers for the next burial. (*see picture*)

After going up the ***"Valley of Blacksmiths"*** we arrived at a place near a spring with very clear water, we decided to camp here and Dr. Fathi and Mahmeed went up the slopes. Very soon we heard their weapon bark; they returned with two dead goats, each one carried on their back as if they were carrying a back pack. They bind right fore and right hind leg together and the same with left, then they pass their arms through them and toss it on their shoulder. Immediately our cook cleaned both animals then hung them on the tree in the middle of our camp.

Blacksmith workshop, in the sun, and under the black tent

Our party stopped every day at around midday under the big oak trees, which were all over the area, to rest until about four or five in the afternoon when the sun was no longer burning.

We camped that night at the foot of a high mountain. Several camp fires were built when Mahmeed asked Blaine to give him four shotgun shells. Blaine curiously asked him “why”? Mahmeed explained with his broken English "there are many thieves and robbers in this area; they already saw us arriving and camping here. To scare them off, I have to shoot in the air at several intervals to let them know we are armed and we will keep one person on night shift every two or three hours”.

Then he shot into the air to let the robbers know that they are not welcome in our camp.

Travelling in wild country is always risky and most of the Bakhtyaris carry muzzle loader rifles of the18th and 19th century. Since the reign of Nasser al Din Shah of the Qajar dynasty, they have smuggled them in to Iran via the Persian Gulf. The British Government tried to stop smuggling of arms for many years, patrolling in the Persian Gulf. It cost them many lives and much effort until they could completely put an end to it.

Our trail up from the river. Can you see the horses on the trail?

The next day we started at dawn climbing to 6000 feet elevation and stopped at noon to rest and let the animals graze near a clear spring coming out of the side of the mountain. This spring had been running as long as anyone in that area could remember. The water was clear and soft, hardly containing any minerals. One could not stop drinking it, the water tasted so fresh and good. Blaine said that was due to the fact that the weather being so warm and dry we had no more fresh water in our water bags.

Chukar Partridges

Mahmeed with two helpers went to shoot partridges for our lunch; others prepared a comfortable place under a big oak tree for us to rest. Soon afterwards he returned with a handful of quail and partridges, we started the barbecue on a wood fire, Blaine was enjoying himself as far as I could see and he explained, "This has always been my desire, to travel along this old ancient trail".

We start climbing to the top (see the trail on the side of the mountain)

Masoud waves while reaches the summit. Note the piles of rock (cairns) where the Bakhtyaris have made prayers when they could first see the place of Sultan Ebrahim Shrine

We carried live chickens with us for a mobile (jujeh kabob); chicken barbeque

Chapter 13

Mahmeed The Bear Killer.

Mahmeed was a great hunter and tracker; he previously worked in a "Zur Khaneh" (the Power House) which is a very old, ancient and historical institution which was a common place of gathering for those who wished to show their physical power or practice building their body. Zur Khaneh was a place like an amphitheater with benches all around the ring, looking down into the circular flat area in the middle of a hall where powerful bodies practicing on all sorts of exercise equipment.

These powerful men in Zur Khaneh were called Pahlavan, "Powerful Champion", or "Powerful Heroic" and they used to have ranks according to their experience and power.

A man beats his drum on one side of the circle sitting at a higher place, singing and reading from tales of fascinating ancient tales and poems. While (Shah Nameh) Ferdowsi is beating his drum all the Pahlavans are in action, each change of rhythm going with a special type of exercise.

The Pahlavans stay in a circle and one by one they come to the centre and toss their heavy massive wooden club up in the air and catch it before it reaches the ground.

After each exercise he goes out of the ring and wraps himself in a red coloured thin towel called "Long", typically hand woven in Iran for use in Bath houses and in Zur Khaneh.
At the end of the exercises they begin to wrestle and the winner would stay and wrestle with the next person until he is beaten or no one else is left to wrestle with him.

While spectators are coming and going they would serve them Chai, (Persian Tea) in special glass cups which are about half the size of normal cups. When they walk away there is a wide mouthed bowl into which visitors can donate their money.

Mahmeed was also from such athletic background. Once he fought and killed a brown bear with only his hands; He was caught by surprise while he was hunting partridges. He found himself a concealed place (a blind) at the entrance of a small cave opposite the spring where all the birds would come for water at dawn. He arrived there earlier than first light, and sat leaning with his back to the rock, and his loaded shotgun beside him, waiting for the dawn to appear. He fell asleep.

A brown bear used to spend his day asleep in this cave and as he was entering the cave realized that someone else was sharing the place with him that day. With an angry roar he woke up Mahmeed in surprise. He went for his shotgun but it was already too late and the brown bear was upon him with no time to do anything with a gun. The bear gave Mahmeed a hard slap across his side so that he lost his grip on the shotgun which was thrown out of his hands and rolled five yards away.

He had no other choice but to fight with the bear empty handed. His intention was to survive and run away if he had a chance to get out of the cave. Sometimes people in stress cannot decide what they have to do in such cases, but Mahmeed thought it would be better to draw the bear outside to have more room and gain a possible chance to fight him, moreover the other friends sitting in the opposite blind could hear them and possibly come to help.

He jumped up on the shoulders of the bear holding his throat. He squeezed, choking him and began pushing his way outside the cave, but there was no comparison in strength. By the time he was in the clear Mahmeed was covered in blood and a hand full of flesh near his shoulder blade was stripped off by the bear's claws.

It was a matter of life and death and Mahmeed could not go on for a long time with such heavy injuries. He was looking around for something to help protect him with and fight, and suddenly he saw a piece of log that he could use, he picked up the log but the bear did not give him a chance to use it as a club, and rushed upon Mahmeed.

He fell to the ground but Mahmeed did not let go of the log and as soon as the bear closed up for the last attack, he got up and jumped to the bear's back, holding the log under his throat and pulling with all his strength. He succeeded in choking the bear, which started to jump up and down and shake himself to get rid of Mahmeed, but he was not the one to let loose of his only chance, going down sometimes and coming up the bear's back. Mahmeed held his tight grip on the log with his knee between the bear's shoulders; he began to feel the bear's movement slowing.

His friends sitting in a blind on the other side of the valley heard Mahmeed shouting and the noise of struggling so they ran towards the place where he was, but by the time they reached him there the bear was going down due to lack of oxygen and there was nothing that they could do to help. No shot gun could be used with the two of them locked together. They were running around and calling to Mahmeed to be careful and not let the log go. They also tried with their hand sticks to hit the bear on his back and his head. By now they could see that each time the bear went down it would rise up with increasing difficulty, each time slower than before. Mahmeed kept his grip tight until the bear was completely still and fell down with Mahmeed beside him.

It was day light by then and they could see that Mahmeed needed urgent medical help otherwise he would not make it to the next day. They borrowed a fast mule from the village nearby and wasted no time in bringing Mahmeed to the place where they had left their Land Rover. Two hours later Mahmeed was attended to by Oil Company hospital doctors, who said there was no hope to see Mahmeed make it to the next day.

He had lost lots of blood from his shoulder blade injury and from several other deep gouges on his body which were bleeding copiously. Moreover infection was now in his blood and he would be very lucky if he should be able to say farewell to his family. But Mahmeed was not a man to give up so soon, he was much stronger than the doctors could imagine.

It was not the first time that he was hospitalized for such injuries. He was also once attacked by two hungry Hyenas during a hunting trip. He ended up with lots of injury and bleeding that took him months to recover from, but he killed both of them with just a stick in his hand.

Blaine said that Mahmeed's manner, and appearance reminded him of his own father, and that was what initially drew them close together. The two always had great respect for each other, even though Mahmeed had little education, and from an outsider's point of view, there would have been little to compare the two.

Late in the afternoon we started our climb up the Zagros Mountains, the narrow trail winding around the mountains and everybody walking now to rest the mules a bit. Suddenly Dr. Fathi saw above the cliff not more than 200 yards away a herd of Ibex looking down at us. When we reached a bend in the trail he took Mahmeed and a helper and from there they climbed the mountain; he was not gone more than 45 minutes when we heard his rifle crack and immediately a second shot, followed by two more.

We had not travelled more than a mile before Dr. Fathi with Mahmeed and the helpers were seen each carrying a carcass on their shoulders descending from the mountain not far from where we were waiting for them. Dr. Fathi came in and dropped the carcass and said "here is your dinner for tonight". We had ample meat to last us for three or four days, all the helpers and boys cheered and immediately packed the carcasses on the mules and then we continued on our trail.

We travelled until it was dusk, when Blaine said to me that we had better stop soon where we can find a place to camp. He was discussing the matter with our group and while I was walking about 10 yards ahead of the others, suddenly I heard something which dropped nearby in front on the trail. I stopped abruptly as due to darkness I was unable to distinguish anything properly, until the object started to rise up from the ground and started to walk on its two hind legs! Then I realized that it was a big Brown Bear searching for berries on the trees on both sides of the trail. The bear was in no hurry and walking softly and groaning he started toward me, and I was so afraid that I began to walk backward not taking my eyes off the silhouette of the bear walking on his hind legs with his arms apart in the air.

I completely forgot that I was carrying my rifle, in any event I could not use it, but in an instant the rear of my foot hit a rock in the middle of the trail and before I knew it, I was lying flat on the ground on my back. Mahmeed from behind saw what was happening and ran next to me while I was struggling to get up and also trying to get the rifle to my shoulder.

Mahmeed said in a low voice, "Please do not move, and stay where you are".

These panicked movements did not change the intention of the brown bear that was walking slowly on his two hind legs towards me. I heard Mahmeed calling everybody not to shoot and also not to move. He turned to face the bear, held his shotgun above his head and started to shout and yell at the bear. The bear stopped advancing and started to roar a few times in reply but when he found that there was no harm from the travelers, the bear dropped down onto his front paws and went walking down the cliffs.

One of the boys started running to the edge where the bear had disappeared, but Mahmeed stopped him immediately and said "only a fool is going after an angry bear when one cannot see more than 5 yards ahead, and if the bear changed his mind to come back it would be your last night and you won't see daylight again". It was pitch-dark when we arrived in a flat area big enough to set up our camp that night. Mahmeed ordered the cook to prepare the dinner and we started to tidy up the gear for next day's ride on the trail to the head of the Karun River, which was about 30 miles away.

Came daylight we saddled the mules and rode away on the trail, high up with a sheer drop of about 1500- 2000 feet on our right, we had a horse loaded with gear and foodstuffs walking in the middle of the file. Suddenly the horse's foot slipped downhill and he fell, sliding and rolling down the cliff about 15 yards. Fortunately there was a small flat place where he landed which was wide enough for the horse to stop itself going down the sheer cliff. We sent someone to recover the pack which was still tied on the back of the horse. They got part of the pack off, but the rest worked loose and fell down the cliff and was not recoverable. The horse was unable to continue the trip and the owner returned home with him.

Blaine explained to Masoud and Dr. Fathi that a horse has his eyes on the front of his head, and a donkey has his eyes on the side. That way a donkey and the hybrid of horse and donkey, the mule, also has this visual advantage. A mule or donkey can see the place where it puts both front and hind feet without swinging its' head from side to side. That is what gives them such an advantage when traveling in rough country.

When Blaine was explaining this the Bakhtyaris were amazed. They had never expected that a foreigner would possess such an understanding of their daily life.

It was only a convenience that the horse wrangler took his horse and went home. He didn't want to be held responsible for more problems and delays. He had only come along because he wanted to see what was happening, and to earn a few Rials which are in such short supply to the nomads.

We were all glad to see the horse and wrangler go without any big argument.

Our trail winding on the side of the mountain

The more we climbed the rougher it became

We used mostly mules, not horses, because of this very problem. Horses could carry more, but were only occasioned on the plains, not in the rough mountain terrain. There is also a lot of difference in the way their fetlock (ankle) is constructed. The horse has much more of a developed arch to their foot above the hoof. That results in them being much "easier riding" than a mule. The mule's steps each end up feeling like they are connected directly to the rider's backbone. After a day riding on a mule, both rider and ridden have sore backs.

We had to go a long way on the side of the mountain to reach the river which we needed to cross.

The Hard Trail to the river. Everyone is tired! You can see Masoud and Dr. Fathi patiently riding behind our wrangler.

The headwaters of the Karun River. Just down from this spring, we were to cross the river.

The next day at noon we reached the Karun River crossing point. We camped there for a day to rest and prepare to cross the river by cable, but we had to send someone to call the cable man, who lived about 15 miles away, in a black tent.

We did not waste time; we needed to bathe, but to go in that fast running river would have been foolish. We tied ropes to each man's waist and the other end to an Oak tree on the bank of the river and dived in the water. No one could stay long in the freezing water which was coming from melting snow above us just a few miles away. We also wanted to fish in the river but the river was very fast and very little could be achieved. Local Bakhtyaris had other ways of catching their fish; a fishhook without any bait was thrown in the river, and they held the line in their hands waiting for a slight movement. As soon as they felt any movement on the line they would pull the line hard. Two out of ten times a Trout of 3 to 5 pounds was caught.

Blaine was very surprised at such way of fishing which was not common to us. For lunch and dinner we had ample fish and of course bread was cooked on the spot and rice was cooked with Mahmeed's supervision. At daybreak the cable man arrived and first we transferred all the gear to the other side and then two persons at a time crossed the river on the cable.

We arrived at the river side the cable is there, but no pulley.
It's time for a swim!

The Mules pushed through the ice cold rapid water. The animals ended up 300 yards downstream trying to swim across 50 yards of river!

Ahooie started first to cross without help but it took him a long time

The cable man pulls Masoud to the other side of the river

This photo taken years earlier, shows how long the same system has been in use.

Resting in the afternoon after a hard crossing of the river.

The mules were pushed in the river to swim across. The poor animals did not appreciate the freezing water and swam hard to the other side. The river ran so fast that they were swept downstream several hundred yards before coming to the other side. The reluctant swimmers which returned were pushed into the water again to swim across. On the other side of the river, we organized the re-packing of our gear. By this time it was midday and there was no point in travelling during that warm weather, so everyone rested and lay down under the oak trees.

From the river we had to go up from the valley through the second range of mountains.
Here is Masoud on his Mule

Ahooie on his Mule

Dr. Fat:.: negotiating difficult trail

Ahooie and Masoud making some easier trail.
The unloaded mule has a bad saddle sore from not using a saddle blanket.

Late in the afternoon we started climbing the Zagros Mountains again. It was a hard climb and in some places it was impossible to ride on the mule and we had to walk the animals. We reached a big white canyon at an altitude of about 8000 feet on the Zagros. It was about a mile long with patches of water here and there with straight rocky cliff walls on both sides. This canyon reflected the heat like a furnace, and we had to be there very late in the day, or very early in the morning to go through.

We decided to sleep through the night and go through "white canyon" early the next morning, before the sun heated it up. By 9:00 in the morning, it felt like an oven.

In the bottom of the canyon in some places the water was only a few feet deep, but to cross it we had to strip off our clothing or be faced with walking in wet boots.

The trail passed through the canyon full of shallow water

Dr. Fathi taking a ride on a Bakhtiyari's back through the water.

On the right foreground, you can see the shadow of the western hat which caused so much interest during the entire trip.

Blaine enjoys his embarrassing ride on this little man's back.

Reaching the other side

We were in no mood to take our clothes off after the hard climb; therefore Mahmeed asked one of the workers to carry us on his back to the other side of the water. This would happen 4 or 5 times in about 100 yards distance, but as soon as Blaine saw the tiny skinny man he said "No way; he is going to drop us in the middle of the pond". Mahmeed said the little fellow could carry one of the mules if we paid him a little money extra!

Blaine, who was still doubtful, asked me to go piggy-back riding first because I was much lighter than he was. It was as if the man was carrying a baby, he had no trouble at all getting to the other side of the water which was about 100 yards away, and returned to bring the next person. Blaine, who still had his doubts, made the ride safely. He was very pleased with the man and could not believe such a skinny man could carry 200 pounds on his back for a hundred yards without stopping for a break, or getting our boots wet.

Chapter 14

Ab-e-Ulma

From the canyon we climbed to nearly 9000 feet where we reached a spring of water streaming out of a mountain called ***Ab-e-Ulma;***, no one knows where the name came from or what it refers to, but as a guess we think that a German traveler (in the Persian language called Almani) passed this way and stopped at this spring.

I must add that there were many fresh water springs running here and there in the Zagros Mountains where the trail passed by. None were as clear and light as ***Ab-e-Ulma***, which was tasteful to all who sampled it. The place around the spring was large and flat enough for many families to camp there during the seasonal tribal movements. We decided to camp there for the night; there was plenty of pasture for our mounts and it was a beautiful place for camping.

What makes a good camping place? Somewhere level to sleep, and a tree to hang edibles.

The night was clear and wonderfully still, the darkness was not yet deep but we could clearly see the stars and Milky Way as if they were just above us on the roof top of a very high building. They seemed so close.

Blaine and Harry and the boys unhitched the gear and set it down. Each person started their daily task of preparing their bedding, the cook started to bake bread and Mahmeed with a helper was busy on the barbecue fire.

After dinner no one was interested in staying around the fire as we did all other nights. We were so tired that as soon as we rolled into our sleeping bags I believe everybody fell asleep immediately in no time. We were awakened up at four in the morning, the sky still full of stars and dark, but we had to start early to be able to pass through ***Tang-e-Mouri***.

Tange Mouri (Mouri Canyon) before nightfall.

Another picture of Tang-e-Mouri

Before we entered this canyon Mahmeed said you must keep your rifles ready here because there will be lots of Ibex on both sides of it. We did not cover more than 300 yards when ***Cherag Ali*** one of the helpers pointed out to Blaine several ibex sitting in the shade relaxed and looking at by-passers. The distance was not more than 100 yards and the face of the canyon was clear for a long distance to left and right.

Blaine immediately asked Dr. Fathi to aim at the two ibex on the extreme right, and Masoud the middle two. He himself would take the three on the left side of the cliff.

We started at the same time and down came the first three Ibex, Dr. Fathi's second and Blaine's second were shot dead and remained between the rocks but Masoud did not have a chance for his second shot. Mahmeed also got his there so in total we had six good size Ibex.

Masoud with an Ibex which he shot for us to eat.

The boys ran up the canyon and recovered the ones remaining between the rocks and brought them down. We had always two unloaded mules that did not carry any loads until they were used for carrying the packs from the tired mules to the fresh ones. Immediately after dressing the Ibex, they packed six carcasses onto the two unloaded mules and we quickly set to go immediately out of ***Mouri Canyon*** before dark because the trail was a very difficult one and you could not see where you walked in the pitch darkness. This was done outside of the normal habit, which was to barbecue some fresh liver immediately after dressing the animals.

We were just in time when we got to the other side of the canyon and here in a wide area we erected our camp for the night. Mahmeed with the boys hung the carcasses on the trees in the middle of the camp and Dr. Fathi did not waste time in preparation of the barbecue fire.

We had a wonderful dinner of all kinds of meat, from goats to quail and partridge. Fresh water came from the ***Ab-e-Bazoft*** which was not more than 200 yards away from where we had camped.

Ab-e-Bazoft from long distance

Everyone was tired and the beds were made for us. After dinner we all went to our sleeping bags, but Mahmeed and Haddad kept the first watch for two hours and then Cheragh Ali with another boy the second watch. I think it was about midnight that we heard lots of shouting and the boys running here and there. We got up to ask one of the boys near us what is happening here, the boy said that a robber had walked in the camp and wanted to steal one of the Ibex that was hanging from the tree.

They brought the man into the camp center, one of the boys gave him few hard slaps on his face and his chest but the man did not say a word. Blaine got up and went to the man and stopped the boy from slapping the poor naked man. We asked what had happened. Mahmeed said that Cheragh Ali is very sensitive when he passes through the land of Mouri clans, because Mouri clans are famous between Bakhtyari for robberies from friends or others. Therefore after midnight when he gave the watch to the others he was not able to sleep after he went to his bed. Cherag Ali while on his bed saw a silhouette of a naked man dragging himself on the ground on his stomach towards the furthest carcass of the Ibex.

He had not touched the Ibex before Cheragh Ali jumped up and seized his wrist calling the rest for help. The poor man was not a lucky Mouri robber and all the boys jumped on him and bound him and brought him into the middle of the camp. Blaine asked Mahmeed to question him, he replied that I have 6 children and tonight we do not have any food for them. One of the boys with a stick in his hand wanted to beat him but Blaine got hold of his stick and would not let him do it. The boy said "he is a liar, they say this is their land and they can steal anything that they come across." They kept the man until day light showed, and after the breakfast Blaine and Dr. Fathi both went to a carcass of an Ibex and carried it to the man and they asked Cheragh Ali to set him free and let him take a carcass for his family. The man before he left went to Dr. Fathi and wanted to kiss his hand but Dr. Fathi did not let him, and told him you can take this Ibex for your family. In the morning we started repacking our gear and got ready to cross the **Bazoft River**, but we had to ride a distance of about a mile parallel to the river to be able to find the shallow crossing point.

Masoud riding alongside of the Bazoft River

As I look at this picture of my best friend riding so peacefully down a lovely river, I wish that I could turn the clock back to better, more peaceful time.

Masoud crossing Bazoft River slowly at sundown

Reaching the other side of the Bazoft River

We then easily crossed the Bazoft River without having to unload our steeds and then from this point we had to climb to a higher altitude again.

As soon as we were on the road again and from this point there was only a narrow trail in which only two persons could walk abreast, and we had to walk the mounts for quite a distance. The cliff was a sheer granite cliff with about 800 to 1000 feet clean drop.

Our next move was to pass through the ***Seyedan land. (Seyeds believe themselves to be descendants of the Prophet Mohammad*)** According to Bakhtyaris they were not hospitable people and very hard to make friends with. If they could get their hands on you, you would not see the next day's sunshine. In spite of that, with all these stories in mind, we had no other choice than to spend the night on their lands before we reached our final destination on Kuh-e Safid (*White Mountain*). We were especially concerned because we were sure that by now they had heard that an American and a Doctor were in the group.

Our camp was made ready in no time and we still had plenty of meat to prepare for dinner, bread was baked and rice was cooked with curry and barbecued wild meat was there for us to enjoy.

The whole night our team was alert. Two persons took watch every three hours but no incident happened that night. At daybreak a man slowly walked into our camp sat far away from us and when boys asked him to join us for breakfast he did not accept, he started to ask us questions, what is the purpose of your visit, how long are you going to stay here, who are these people with you, and what are their ranks and so on. After he found out that we are only passing through he left us without saying goodbye.

The faithful Mahmeed on foot beside Masoud on his mule.

We continued our journey, again starting to climb to higher altitudes. We arrived at a large green meadow with lots of trees which surrounded a lake about half mile long. The name of this place was ***"Shalamzar"*** a place that all tribes must pass on their semi-annual migration. This was truly a beautiful green area, around which they would camp before they continued their long and hard trek to their summer and winter quarters **(Yilaq and Qeshlaq)** back and forth.

Mahmeed always walked beside Masoud, a loyal friend protecting him

The place was situated in a flat area but surrounded by high mountains. Here we saw two shepherds camping on the edge of meadow. We sent someone to talk to them. They told us that they had seen, early in the morning, several Ibex with long horns crossing to the opposite mountain. To hunt now, meant that we would have to stay overnight in that beautiful green place which was home of every tribe passing to go through the Zagros mountain. We did not waste time, Mahmeed ordered the men to erect the camp ready for night and prepare food, and to cook bread ready for our return. We took two boys to help carry water and our packs; we hit the road and left for the opposite mountain which the shepherds had shown us.

We were not more than half the way up when Mahmeed pointed to our left. He had spotted two young Ibex. Mahmeed said he believed that there were more there. He had amazing vision, what Mahmeed could see with his unaided eyes, we required a binocular to see. He led the way nearer to the game. Every now and then he would stop and check the wind direction.

The wind in the mountains changes to a different direction depending on the structure of the land, therefore if you wish to hunt on high altitudes you are bound to check the wind direction to be on the safe side. We had only an hour and half until sunset, and delay meant no meat for the camp. Finally he pointed to a ditch. He asked us to lie there and wait. He crept up a hump and from between the dried bushes he looked straight at the opposite side of a canyon. He knew exactly where the game was, with his hand he beckoned us to climb up near him.

We proceeded slowly and carefully, and within a range of 150 yards we saw them there. Mahmeed was so experienced with the terrain that he could take you as close as a few yards of the animals. In the past we had asked him not to do it anymore as it was difficult to shoot with scoped rifles when the game fills the entire field of vision. First Mahmeed raised his head between the bushes and said there are more than twenty of them here. The majority were Ibex and he showed us which ones were best to be shot for tonight's barbecue. Blaine had to start first on the left one and I took the right side Ibex. As soon as his rifle boomed I shot my first one at about 120 yards, quickly moving and taking the second one again. \next I heard Blaine and Dr. Fathi both shoot theirs and then later on a shot echoed in every direction in the mountain, we could not exactly say where from, but after some time we saw Mahmeed appearing from behind the ridge carrying a good size Ibex on his shoulder walking leisurely towards us.

It appeared that before we started shooting that our Mahmeed had disappeared behind the ridge and had sat there waiting. Mahmeed knew exactly in which direction the Ibex would run after we opened fire. We ended up with five Ibex. To take them down the mountain by ourselves would not be an easy task. We needed to have help from the camp. That turned out to be easy. The Bakhtyaris have special calling signals which they have maintained from medieval times. They can easily understand each other from a long distance apart. Using their signals they quickly asked for two Mules to be brought for carrying the game.

The boys started to clean the animals and the best tenderloin parts were kept for us to be cooked over charcoal. Mahmeed also washed the livers and cut them in small pieces and wrapped each piece in fat tissue which came from the belly of Ibex. These we put on sticks over charcoal. This was an appetizer which was ready before dinner. What a delicacy, you should try this recipe to barbecue the liver of the wild game the way described above. You don't know what you are missing if you have never tried it. When all returned to camp, Blaine said "What luck, we can now bring some of these for the Khan as well." You could not imagine the temperature change at such high altitude. It drops drastically as soon as the sun disappears. Everybody had a wonderful dinner with lots of butter and ***Doogh*** (*Yoghurt drink*), which came from the shepherds camp. They were very happy to receive meat in exchange for the ***doogh.***

Sometimes we have been offered to trade several tame sheep or goats in place of an Ibex. I did not know why Bakhtyaris are so keen to have Ibex or wild game until Blaine and I started to question them to find that they believe in the Ibex flesh as medicine and curing much sickness which they could not find in any other meat. They believed that they received the strength of the wild game by eating its flesh.

Dinner preparations. Meat is being kept warm over a pot of cooking rice.

Soon we were near Kuh-e Safid *(White Mountain)* where we were going to stay for a while and hunt. Black tents were pitched all over the White Mountain. We now were descending from the higher trail towards the black tents and we could see from a far distance their settlements. Some men were running towards us with bowls of *Doogh* (*Yogurt drink*). This is a custom of hospitality with the Bakhtyari people. This was a special *Doogh*, with a sprinkle of mint leaves and ice in the bowl. The first question we asked was; where do these people get their ice. The answer was that the ice came from last winter's snow. It was gathered in a hollow in the ground or in an ice cave deep enough to preserve the ice for the whole summer. It was lovely drink after such a long journey. This was the first time that Blaine had experienced Bakhtyari hospitality and their unique nomadic places, where no westerner had set foot before.

Wild Zagros Mountains, where the geology stands on end!

Chapter 15

Kuh-e-Safid (The White Mountain)

We finally reached the summit of Kuh-e Safid plateau, this mountain is very big. We could see clearly many black tents from large to small depending on the size of the family and the wealth of the person who lived in it. Walking and wandering between the nomad's tents towards the Khan's tent the nomads clustered about us and women with their colourful Bakhtyari clothes. The women wore long pleated flowery patterned skirts and shirts, with bright colour scarves. Children were laughing and running about bare foot, talking among themselves and from time to time saying "Khosh Amadie" (*welcome or glad you came*) with a shy face. Sometimes after the word they would hide their faces in their scarves.

Kuh-e Safid in spring

The shepherds were coming down the mountain carrying their new born lambs on their shoulders, proud and happy to be back in camp. Dogs were treated as pets but slept outside the tent and they were fed after the meal was over. Some were vicious and barked hard but if you went down as if to pick up a stone they would run away and not turn back. Once you enter their tent they would not bother to bark at you anymore.

The nomads were curious about our visit but did not seem to mind our presence. The women with babies on their arms come forward to have a closer look at the strangers, especially the tall American visitor, which for the first time in their history an American has set foot on their soil. Blaine enjoyed these fascinating sights and with all the grace he could muster, and a smile on his face greeted them with the normal, "Salam" (*hello in Persian*), which was met with astonishment, especially among the women.

The women started to ***"KEL"*** (it is a sort of joy thrilling sound they make from behind their throat in a high tone) The Khan had already arranged for ***Tushmal*** to play ***Saz-o Dohol*** Bakhtyari music and the women started to dance (***Dastmal Bazi***) ***(The Kerchie dance)*** and in no time the men also joined the women and started to dance together. What a welcome reception, everyone seemed to be happy to see us.

Black tents are found everywhere in the Zagros Plain in the summer

Yushmal or Saz–o-Dohol welcoming our arrival

Bakhtyari men started (Choob Bazi), or "stick play dance"

Dastmal Bazi (Kerchief dance)

Bakhtiyari women all came out of their tents.

With their children carried daily on their back

Milking the goat

Mahmeed showed us a place where we should erect our tents it was between few black tents and also convenient flat surface, close to the spring as well. Blaine told Mahmeed that it is proper to pay our respect to the Khan first before we settle down to our tents.

We were directed to the tent of our host ***Shir Ali Khan Shehni*** head of the tribe in area of Kuh-e Safid, a distant relative of our guides, Mahmeed and Ahooie. The Khan came out of his tent and walked fast towards us and first he greeted Ahooie and Mahmeed then shake hands with us and saying ***Khosh Amadid*** then he directed us to his large black tent, he was very happy to see us and he offered us make ourselves feel at home and asked Blaine to sit beside him on beautiful Bakhtyari carpets with round cushions set beside each other all around where the carpets were, after welcome tea and ***Doogh*** was served.

Meanwhile Blaine presented to him two carcasses of Boz-o Pazan (male and female Ibex) that we had shot in the morning on our way to him, and then Khan called the boys to hang the goats on the trees and clean them and immediately distribute them amongst the settlers.

Without refrigeration, in the summertime temperatures, meat was not allowed to spoil, but was quickly shared by all. This tradition was repeated over and over everywhere we went, and I suspect that the temperature was the primary reason for the sharing rather than Taarof.

Outside ***Tushmal*** was playing their ***Karna va Dohol*** (Trumpet and drums) and dancing went on for few hours until dinner was served.

The Khan ordered his wife or wives, (we couldn't tell how many wives he had and it was not considered courtesy to ask such personal questions), and his Monshy ***(head of the servants)***, to kill a young goat and some chickens for supper and he added for my honoured guests they were also ordered to cook ***Pillow (Rice)*** for dinner.

Preparation started and you could see that they were running from one side of the tents to the other, and to the spring of water, which was not very far from where we sat. All were busy doing their utmost for our comfort. Dr. Fathi set the small electric Honda generator that we brought with us up and extended the wiring from our tent to the tent of the Khan and the tent of the cook plus a line to the spring.

Our camp on the top of Kuh-e-Safid. Note the Honda generator on the lower right.

This little girl stole my heart. Her job was to carry water around to everyone in camp. Because I always had something sweet when she came to deliver water to our tent, we always had cool water waiting for us. If you look carefully you can see the water bag on her back, which is made from a sheepskin. She is standing in front of the watering trough where we put the electric light.

Unfortunately after all our good intentions, someone came to let us know that the animals refused to go for water while the lights were burning near there, because they were not used to the electric lights, and some people in the village had not seen them before on these remote mountains, we had to turn off the lights to the spring for the animals would return to drink water.

After we had erected our tents and washed and rest for a little while we had to go to the tent of the ***Shir Ali Khan***. Talk started, first with the Khan questioning us about our job, positions and ranks. After we gave him satisfactory replies, then discussions opened as to how we should begin our hunt in the morning. While we were sitting there, men dropped in from adjoining tents and further places opposite mountains and some from 3 to 5 miles distance near Kuh-e Safid (White Mountain).

These are customs born in antiquity, within the Bakhtyari nomads, if they have special guest for dinner other adjoining residents or new arrivals can come without being invited to join the dinner with the guests, the hosts servants always prepare more food than it is required for such incidents. When we brought the Ibex, we were contributing to this welcome dinner, in the same way someone would bring flowers, a special desert, or a bottle of wine to a dinner party in the west.

We were sitting in a circle on the ground and all joined in the conversation; many of them knew the land and giving their idea. Those who were best in tracking were asking to accompany us in the morning expedition, between them the talks continued until midnight. Of course we were not able to decide where we should begin unless they could come up with their suggestions. After all we had to ask Ahooie and Mahmeed to program the next day's trip. Only they were able to accurately determine which person was simply "putting their best foot forward", and which were actually knowledgeable about the hunting. The locals were of course, completely in the picture about who knew what, but they would never dream to tell us, and possibly hurt someone's feelings. They had to live with their fellow citizens, and were probably related to them anyway.

Dinner was ready they brought a long white sheet laid in the middle of the tent on the carpet and plates for everyone but no spoons or fork and knives, vegetables from their gardens such as spring onions and radish was served and someone brought a sort of jug of water with a bowl and a tiny small piece of soap that everyone can wash one hand according to custom.

Dinner was served in a large round tray of about 45 inches diameter with saffron flavored rice and the young goat in whole cooked and set between the rice, several other dishes of stew and vegetables, raw onions, spring onions and lots of dates, tomatoes and ***Nan Tiri "Tiri Bread".*** This bread is made very thin and prepared on outside curved cooking plate on open fire and it is rather delicious. Masoud asked for spoon and fork and Khan immediately ordered to fetch them for us. It made it much easier indeed to eat with spoon rather than with your hand. After we had our dinner again they brought the jug and dish to wash your hand and then tea served several times.

After everybody sat back, and the Khan turned to Dr. Fathi and said, "we have a few people within our tribe that they need medical attention and we are not able to send them to Esfahan or Masjid-e Sulaiman. While the good Dr. was sent to us by God we request Dr. to look at them and see what he can do for them. As they knew a Dr. is arriving here they are now waiting outside to see him. For next few hours Dr. Fathi was busy consulting and distributing medicines to the settlers of Kuh-e Safid. Thanks to my wife's advice to take jars of Vaseline and crèmes with us for the women to use on children and babies bottom with great satisfaction.

They brought a young boy from nearby settlement that boy had fallen from a rock on his knee a month earlier he had wounded himself but while there was no hospital nearby, their medicine man or old wise man of the compound put ***Dew*** on the fresh wound and wrapped it up with a dirty cloth, it was wrapped more than two weeks since we opened it. The wound was infected badly and the boy could not walk, his knee was bending and with great difficulty and lots of pain he used to move around.

When Dr. Fathi opened the cloth it was as if the wound was caused by shrapnel, it was white and red and you could see part of the white knee bone under all dirty flesh and skin hanging loosely around the knee cap. Dr. Fathi, very angry and furious, he started to work on the boy's knee washing the wound with antiseptic liquid that he had with him gave him a 1.200.000 unit penicillin injection and instructed the father of the boy not to cover the wound until completely healed and recovered, Dr. Fathi ask them to bring the boy to him same time every day for his antibiotic injection.

As we were planned to take off for hunting next day the father of the boy said that he knows where we are going to stay at night so he can bring the boy on a mule to us. Indeed the next day and days after father of the boy followed us and brought the boy on mule into our camp where we were going to stay for night.

Not all the medications were western in nature. We also brought with us cans of instant coffee, and many small packets of individual servings of instant coffee. These were given to people who had all sorts of stomach ailments. This coffee was not brewed, but taken as a spoonful of powder. The individual packets were given so they could keep track of it, and keep it from spoiling. It was a powerful drug among them.

We were shown to a black tent which was erected for us from days before we arrived, it was out of question to talk to them that we wish to sleep in our own tent, was not done in hospitable Bakhtyari tribal way, but we used our own sleeping bags. Early in the morning before light the shepherds were already directing their herds to the spring and then each flock were going their own way. It was before sunrise that we were called for breakfast Eggs, honey and warm fresh bread was prepared and finally with lots of ***Doogh***.

All of us were ready in half hour and started to go down the mountain to the bottom of the of canyon where they have decided night before to take us there first, we had not gone more than 400 yards down that trail, we saw the side of the mountain is moving.

Mahmeed told us to sit 10 minutes here until light comes, when the first morning light showed and our eyes were more used to the twilight we could see that there were so many Ibexes in sight not more than 150 yards away from us under our cliff and on the opposite side of the canyon, Blaine showed Masoud and Dr. Fathi where to sit, and which area to cover for shooting, Blaine himself did not see any species worth of trophy among all the Ibex, but as we were obliged to provide meat for the inhabitants we decided to shoot some for them.

This is truly a "trophy" Ibex. It's body weight would be about 300 to 350 pounds.

We got five Does and a small Ibex between us, they loaded them on one of the mules and sent them back to the settlement for the Khan to distribute to the people there and we continued our journey to bottom of the valley. Everywhere we saw tracks of bear and people with us said they have seen them daily around these mountains, by the time we got to the bottom it was five in the afternoon and we saw several more ibex on the opposite cliffs that they came to lick the salty stone, there, we could use some for dinner that night, we shot two and wounded one that we could not track it in the dark.

It was time to look for a suitable place to lay our bed rolls, in the middle of valley a stream of fresh melted snow water was passing, on one side there was a piece of rock as big as a small house with natural flat top no way to climb on it, the rock was about 15 yards high, about 35 yards wide and 30 deep.

Blaine asked for the rope and he tied up a piece of stone on the end and threw the rope across the rock and from the other side two persons hold it and Blaine climbed up the rock, then pulled everyone up, the Bakhtyaris were completely astonished and no one of them ever tried this before.

The rock was flat and big enough to house 12 persons sleep on it comfortably and there was enough room to make fire for our cook to prepare dinner and make barbeque, after dinner we sat around the camp fire and discuss the next day's plan, they said that there was no one as far as they know who could climb this rock ever before. Some of the boys started to sing their native songs and others were happily clapping their hands with the rhythm

Joyfully everyone went to their bed rolls, It was not midnight yet that someone woke me up asking for the torch that I used to carry I gave him the torch and I followed him to the edge of the rock on the far side, as soon as he fixed the light on the bottom of the rock I saw a big brown bear standing on his hind legs and scratching the rock and making funny noises, by this time all of the boys woke up and they all came to look what was there, after some time the bear left and we went to our sleeping bags and slept deep under the clear sky full of stars.

Morning came and we had to continue our expedition to the next canyon, breakfast was quick and all went down the rope. By this time the mules were brought from the upper valley, because it would have been dangerous to let them loose in this canyon. They loaded them, and suddenly Mahmeed called and said that we are missing one of the Ibex we had shot the night before. Apparently the Bear had managed to grab the carcass which was hanging high on the side of the big rock and took it away.

We started from bottom of canyon to climb to the far side of the White Mountain, by noon time we came across a large canyon and we spotted two Ibex. Masoud shot one and Dr. Harry Fathi the other, we skinned them and salted the skins and send the carcasses with the skin back to village keeping only a leg for our dinner.

Masoud with the Ibex taken in the top of the Zagros.

This time of hard travel afforded some times of fun to relieve the exhaustion and stress. The Bakhtyari people are very much a fun loving people, and show it here. I have learned not to take offense where none is intended!

It was late in the afternoon when we reached a large canyon, there we decided to camp for the night, Mahmeed and one other went for Partridges and we had ample game meat to ourselves. After dinner Blaine asked one by one to tell us their memories of past, this way we had spent the evening and Cheragh Ali used to sing every night for us Bakhtyari songs.

At day break Mahmeed and Ahooie went on top of the canyon to look there for any game, as soon as Mahmeed bend over the wide canyon he spotted a big brown bear 700 -800 yards away to the left side of the canyon. He came back and showed us to the spot, we saw the bear and Blaine said it must be an old brown bear. It was difficult to reach close to the bear easily and to get down there was no immediate access unless you go around the canyon then descent from far side of the mountain. One had to walk about 3 miles before he could decent the cliff.

To be able to get a good shot at the Bear we asked Haddad, who was one of the boys with us, to go around the mountain and find a way behind the bear and shoot few round of shotshells to scare the bear. The idea was to have it pass under where we sat at about 300-400 yards at almost a vertical angle down the hill. Haddad reached the place after an hour of hard walk but the bear smelling the man's approach suddenly went under a big rock where Haddad was approaching it.

From where we sat we tried to call and to warn him but unfortunately the wind was blowing towards us and our calls could not be heard, Haddad reached the stone climbed it and he was looking everywhere to see the bear but the bear crept under the stone far in the dark, we sent someone running towards him to warn him were the bear had hidden. As soon as the man reached a point opposite Haddad, he called and warned him not to go down the rock.

By shooting a round of shotshell the bear took off running on an straight line under us but to be able to shoot we were required to lean down on our stomachs and hang down the rock above the valley. This was not possible without someone holding your feet tight so as not to fall 1000 yards down the canyon. This was precarious to say the least. For the first four shots Blaine held Masoud's legs and he shot the bear twice in his side and in one of his front legs. These rounds took no effect, and the Bear did not reduce his running speed. If anything, he was going stronger, the two other shots were missed, Blaine replaced Masoud and ask him to hold his legs firm, Blaine went flat on his belly and shot the bear 4 times but all bullets hit the body without any serious effect. This shooting was done at a range of near to 500 yards, and the rounds were nearly expended of their energy before impact.

Chapter 16

Blaine

"I was afraid that if the Bear made it to the end of canyon there will be very little we could do to go after him. To chase a wounded Bear would be a suicide business. I feared we had "jumped right out of the frying pan, into the fire". We had succeeded in getting Hadad clear, only to place him in imminent danger again, if the bear returned the way it came. I quickly changed a fresh cartridge into the chamber, and this time I shot much more deliberately. I had also seen what the trajectory was in this vertical position, and had fine tuned my aiming point. I was hanging over the rock with most of my body in the air and aimed, waiting for a good moment and I shot the beast right in the head."

The bear started to roll down the canyon going down for about half mile and falling from a sheer cliff about another half mile directly to the bottom of the canyon. We were not able to see the place where it had fallen. It was about noon time that Dr. Fathi arrived and joined us from other side of the mountain where he was hunting Ibex.

He asked one of the Bakhtyari men how long it will take to get the skin up, and also he wanted the tenderloin part of the bear to be brought to him, no one was interested to go down the vertical canyon, but when we offered the local equivalent of 50 dollars reward three of them agreed to do it. Meantime we moved to a new area and Masoud shot a trophy 48 inch curved antler Ibex which was perfect

Aziz Qoli climbing in the canyon; Picture taken by long telephoto lens. These Bakhtyari men went up and down the rocks like goats.

After the bear rolled and slid down to the bottom of the canyon, the Bakhtyari men brought it back up onto a ledge where they could skin it and bring the skin back up to us.

Night was coming and we searched for a place to rest, and by the time dinner was ready. The three men dispatched down the canyon, arrived and brought the bear skin, we salted the skin with what little salt we had with us and with great difficulty loaded it onto a mule. The poor animal could sniff the bear and was refusing to accept it. After some calming, we sent it back to the village on Kuh-e-Safid.

Strict instruction were given to the boys that as soon as they arrived they should spread the skin in such way to allow maximum air and sunshine, to allow it dry soon. Because the Bakhtyari people are so independent, we assumed they would overcome their Muslim traditions that bear and wild Boer are considered unclean. The skin was left rolled and untouched in the village, no one was willing to spread the hide. By the time we got back the hide was rotted out and the only thing we could do was to throw it away in bottom of the canyon.

Look at the size of the bear's head. It is almost as wide as my hips across his eyes.

Blaine posing in front of the bear skin

This was one very big Bear!

Haddad tired of climbing up the canyon sitting to rest a while

Everybody happily posing in front of the Bear skin.
This was the end of a dangerous hunt for Hadad (between Masoud and Dr. Fathi)

Those paws and claws could do a lot of damage.

Dr. Fathi was very happy to get the tenderloin part of the Bear, he started his preparation for Barbecue and when the dinner was served he served us the bear meat as well. No one of the helpers touched the meat or tried to eat, but we tried and found it excellent and full of spice, a little wild flavored with onion in the background. Masoud and Dr. Fathi thought it was very delicious and nice to experience bear meat for the first time. Blaine had eaten bear many times in the past, and didn't care that much for it. Bear meat in the late summer or autumn is full of fat and tallow. It is like eating old mutton. After eating, it sticks to the roof of your mouth. All the Bakhtyaris were disappointed with our taste and even showed their faces hardened during the time we ate the bear meat.

Our helpers took the bearskin back to the base camp at Kuh-e-Safid, to salt it down, and scrape the fat, so that we could make a rug.

Come daylight, in the morning we continued our journey during descending from a canyon Masoud's shotgun dropped from his shoulder and somehow he wanted to see if the mechanism still working, therefore, he cocked the shotgun and pulled the trigger without any shell in chamber, he thought it works alright, but when Mahmeed went to get morning Partridge and Quail he found out that the firing pin was broken.

"Blaine said later it is always safer to pull the trigger on an empty shell, less chances to break your firing pin".

Blaine aiming a test shot with the Muzzle loader

Blaine asked for a nail, but no one had one, so he looked around and found the cradle box which was on the mule for carrying loose things. He pulled a nail out of the wood and with a hammer which he made from stone he had to work harden the iron into steel using charcoal to increase the carbon content, and then temper the new firing pin. He brought it to the shape he wanted, and fixed it in such way that for another seven years Masoud and Mahmeed used the gun, the pin was strong as new.

This incident caused a delay to our schedule and before we left the camp it was one hour past noon time, we just came over an edge of the canyon that Mahmeed pointed towards far left that he spotted several big Ibex, we all went down on our bellies with binoculars in hand we carefully spotted them near a water hole, three of them were making good trophy with their big antlers.

Dr. Fathi was assigned to right with a young man to carry his bags, Blaine to the left and I in the middle of the area about 200 yards apart from each other. We came back successfully with a trophy each, but Blaine had done better than all of us, he had shot two Ibex we sent all the game on a mule to the Shir Ali Khan on Kuh-e Safid to distribute to their people.

Getting dark we had to stop by this time, it was impossible to carry on in dark without moonlight, we camped on a flat surface right above the edge of the canyon near a clear spring was situated about 40-50 yards down the cliff. The boys made several camp fires around the place where we were going to sleep. The boys said we do not want to take any chances of being surprised with an unwanted intruder.

Come daylight we found our way to go to the next canyon, here we saw many more game animals, suddenly one big Ibex came into our view at good 400 to 500 yards away. Only one good thing was that the game was vertically high above us. All rifle trajectory is based on level shooting, and with high angle shooting, the bullet will hit high over the target. This is the same either uphill or downhill, if you remember this you can adjust your aim.

Masoud aimed with his 7mm Remington Magnum accurately and the animal was hit right on the shoulder blade, dropped and rolling down the cliff, the boys started to cheer and they built a cone shape marker from the stones putting on top of each other, about two yards high with a bottom diameter about one and a half.

Two others climbed up the top of the cliff where the animal was first seen hit, they built the similar marker there, when I asked why they did this, they said this is to show any one who passes here the distance that the animal was shot. No one will destroy these markers. We were sitting having our lunch when we saw two Bakhtyari men came down the opposite mountain, they were friends of the people we had with us. One of them said their food was finished yesterday and that they lived on berries and wild fruit since. They needed some food. Mahmeed gave them a front quarter piece of game meat, and some bread.

Mahmeed asked them what they doing here, the man replied that they were hunting but they were not successful because their black powder was stored near the water bags and was not functioning properly, therefore they did not get anything. They had one Muzzleloader gun of late 18th century but they hid it way back before they walked into our camp, because they were not sure who we were and guns with no license is not permitted.

Mahmeed had to convince them through their friends that we are not from military officials, the man returned and after half an hour came back with his long barrel Muzzleloader and Masoud asked me if I wish to try shooting with it. After I made a shot I asked them if they are willing to sell his gun, and they agreed after a hard negotiation, I bought the gun which I still have on the wall in my house.

When we went back to the base camp at Kuh-e-Safid, we found the bearskin laying outside the camp. It hadn't been touched as we had asked. The skin had been laying in the hot sun for three days, without scraping and salting, and had thoroughly spoiled. The hair was falling out, and it was worthless for our purposes. When we asked why they hadn't salted it down, they said that the salt which we had provided was far too valuable to use on a bearskin.

Since we were visibly disappointed, we just walked away, and went looking around the village. We found three women making a carpet. The three women were a mother and her two daughters. One daughter was married, and the other was too young to have been arranged for.

The father of the two daughters had been killed by a bear the year before, and these three were making a carpet to take down to the city in the winter to try to make enough money to live on.

The Young girl who was still unmarried is sitting in the center of this picture. Her mother immediately began to bargain with us to marry her to one of us. As I was the youngest, and a foreigner, she immediately caught up on the idea that I should marry and provide for her daughter. I politely explained to her that I was already married, and had four children, but when she said that I was in Iran, and could have four wives, her daughter being a summer wife, I said that the "Winter Wife" would not be happy, and I offered instead to buy the carpet, unfinished, and that they could deliver it when finished. I paid about $60 US for the carpet, that being a high price, but enough to get me out of the situation, and we left.

Imagine my surprise when Masoud called me the next spring, at my office, and said I should bring Judith and come to his office. The lady, true to our bargain, had brought the carpet to us finished. (I think she wanted to see the "winter wife" and know what was so special about her, that I didn't want another.

We paid her for her trouble and travels, and took the carpet, to a special place in our home, where it still sits under the pictures of the family, and the half-finished carpet.

Here a young Bakhtyari girl makes a carpet from wool welt
In a wooden frame, using hand dyed wool for the warp.

Chapter 17

From Kuh-e Safid To Esfahan

The Bakhtyari lands during the 1970's were a wild and lawless place, where no one was able to travel without permission of the local Khan. The Bakhtyaris used to carry Black Powder Muzzle Loaders with long barrels since 1870 and later on they could obtain modern rifles, they were very good sharp shooters, one person at the top of canyon trail could stop an army column from passing through their land. If they knew, however that you are the guest of one of the Bakhtyari Khans they would treat you with respect and do whatever they could to provide for your comfort.

The ***Tashrifat*** and ***Taarof*** is a code of respect, manners, and formalities between Persians which is difficult to understand or translate, this is the most distinguish characteristic of the Persians, for example when you go someplace and ask the host may I enter, instead of saying ***Biya Dakhel***, which means simply come in, he will use a sentence which would be thought of as flowery sentence "***Tashrif Biyavarid va Ghadam shoma bar Cheshm",*** which means come in please and your steps on my eyes.

It is a strange thing to watch ***Taarof*** being observed. In normal Persian cities and villages, it is a more or less formal procedure. Of course in the Bakhtyari areas they do not consider themselves like other Persians. They have their formal life, and like everyone else in the world they have their informal life, around friends and family. The formal life is often observed in public, although when found in a new, or different situation, they will sometimes fail.

When greeted by a foreigner, they will often forget ***Taarof***, and the situation can become comical. Like everyone else in the world they have their own specific vocabulary which they use on a daily basis, and pronounce with their own accents. It was truly interesting to be introduced to a group of Persians and watch them try to pronounce my name "Taylor". It came out in some very strange ways. The most common was "Tire Lever", which was what they called the iron bar used to break rubber tires from rims. Although we didn't see many rubber tires in the mountains, they mostly addressed me as "Tire Lever".

Most of the time we kept our rifles and shotguns packed away in cases, to keep them from being damaged. At times, however, when traveling in some of the wilder places, or if we needed to push through after dark, we armed ourselves. In one of these wild roads, we were traveling with our weapons lying across our saddles, when we rounded a bend and met an old man on a mule. He was in his sixties or seventies, and had a young teen aged girl walking behind him. She was of course covered from head to toe in the traditional "Chador", meaning tent, and walking a dozen yards behind him on the trail. When we came into sight, and the old man beheld 25 to 30 men, some of whom were armed, he began to wail. At times like these my Persian and the small amount of Bakhtyari language I had would fail me. I couldn't understand a word that he said. I was very uneasy, because I didn't like to be in the dark about what was happening around me.

Old man "(pir-e-mard)" with his young bride of 15 years of age, taken from a long distance

The old man and his young bride make a safe getaway from the gang of riders.

Little did I know however but the rest of the group was just as confused as I was. There was only one person in our group who could understand his dialect. Ahooie who had grown up in the mountains a very long way from where we were recognized his dialect. The old man had travelled down to a larger village, and bought himself a young wife. He was petrified with fear, and sure that we were there to murder him and take his young wife. Only after some few minutes was Ahooie able to convince him that we meant him no harm. After he became more normal, his language reverted to the local Bakhtyari dialect, and others could understand him.

During the time of the explanation, I looked over at the young girl. It was quite obvious that I was very different than the other men, being much lighter, and much taller. I was also dressed so differently, with the Stetson hat, blue jeans, and cartridge belt strapped around my waist. When she was sure nobody else was looking, she dropped her veil, and gave me a toothy grin. I did my best to return the smile and marveled how the old man, who must have had extremely bad eyesight, could even negotiate the trail. This was one very unattractive girl. Later when discussing it in private, Masoud told me that the old man had probably not ever seen his new bride's face, and that she would not reveal herself until the deflowering ceremony.

This behavior of people breaking their very strictly observed taboos in my presence was repeated again and again. It was like they didn't consider me as a foreigner, to be subject to their rules. It continued throughout our travels in the wilderness, and on some occasions was to our great advantage. I often hid from people how much knowledge I really had, and feigned a complete lack of understanding of the language. I heard many things because people thought they could talk around me, without any sense of withholding.

Masoud tired of ridding Mule, walking on the old trail

We travelled two days, in the afternoon of third day we rested near a garden in outskirts of ***Birgun*** village. What a beautiful place. There was a small stream running through some trees, fat tailed sheep were grazing on the grass, and it was probably the most peaceful and beautiful place we encountered on our entire trip. Someone from the local people came to see us and brought a small basket of first class grapes. Dr. Fathi asked for more and paid him a small amount of money to get some more grapes.

Sheep crossing a bridge near the village of Birgun

Dr. Fathi only gave about the equivalent of 75 cents American, but the man dutifully went away and came back later with a donkey. The donkey had two big pannier baskets, one on each side containing grapes. The grape juice was being slowly crushed out of the bottom of the baskets. The man explained that he didn't know if we wanted red grapes, or green grapes, so he brought a basket of each. When we arrived it was about 11:00 in the morning, and we lay around the garden, eating grapes, and made a temporary camp for lunch. Our group ate both of those baskets of grapes. When travelling in the high mountains, you tend to be a bit dehydrated, and we were eating a lot of cheese. Constipation is a constant problem, as long as you drink pure water. Let's just say the grapes did their work that day.

At about 3:00 in the afternoon, we continued our travel until late in the afternoon when we got to outskirts of ***Qale-e Bidemi***. Here before we entered the village a gendarme stopped us and when he saw that we have so many weapons with us he started to say ***Bah Bah*** means ***well well*** in a tone as if he got criminals there, he started to question us and wanted to know from where we come and what was our purpose of travelling in the nomad lands, beside who gave you permission to carry these weapons which are illegal.

Mahmeed tried to convince him that we were not criminals and we had permits for the guns and we are also had a permit for passage through Bakhtyari lands, he said if you further have any questions you can ask Colonel Akhbari.

The word was magic and within seconds the whole atmosphere changed and was replaced by polite talk and the man started to say that you have to understand that I am only doing my duty and nothing else and if you have permits I do not have to waste your time any longer,

Masoud said to him that he must check the licenses if he is doing his duty, but he now tried to be extra nice to us because of his first impression when he saw us, but Masoud did not accept this, and asked Mahmeed to show the licenses in a very formal way. After that Masoud asked him if he was satisfied. He then asked him if he had received a telegram from police/military in Ahwaz in respect to our travel through this land, he said that he must ask his colleagues because he was travelling most of the time working out of his office, and just now he had arrived after ten days travel in the Bakhtyari lands.

This gendarme was treated quite harshly, because he was in the habit of harassing all around him. He used his authority to make himself fat with the bribes he could get out of intimidation due to his office. When called to a higher authority, he folded up like a paper house, and couldn't do enough to make our travels through his area pleasant and ***fast***.

We didn't wait long for the ***"Monshy***" or the secretary of the head of the village ***Ali Mardoun Khan*** arrived and said that the Khan would invite us all for dinner tonight at his ***"Kolbe Naqabel"*** (Cottage or Hut) which is not worthy to your standard. (More Taarof).

Here the Gendarme said goodbye to us and he said he would be at Khan's place to join us for dinner that night then we gathered our gear and we were directed to the Khan's house.

The house was made of mud and straw bricks but very well constructed with several large rooms and a guest room for about 50 guests, It had a big yard and some small rooms at the far corner of the yard for the servants and helpers to rest. We unloaded our stuff in the yard and we gave the Khan a carcass of Ibex that we had shot early in the morning. The Khan had already prepared dinner for us with their local bread which is called ***"Noone Tiri" (Stick Bread)*** very thin bread as thin as 180 gr. per sq. meter, light and very delicious; you can't stop eating "***Noone Tiri***" when served with ***Kebab.***

The Gendarme arrived with one of his colleagues and when he saw Masoud he went to him and wanted to kiss his hand, but Masoud pulled his hand away ***(this is custom in Iran, when someone wants to kiss your you pull your hand away to give him an honour and respect)*** the man said I checked with our post and indeed we have received the telegram of your arrival more than two weeks ago but you have to forgive me if I was not kind because I have strict orders to be careful in here and check all new comers.

Dr. Fathi made some funny remarks and changed the atmosphere of the room in a pleasant way, we all sat down and the Khan started to talk, he said that some three days before we received news that a Farangi together with some Iranian officials would be passing through here. "That is why I set a post outside the village to warn me before you go through unnoticed".

Dinner was served and every one happily filled their empty stomachs, then normal questions started in regards to the ranks of each person and their family size, the Gendarmes said goodbye and went to their homes and the Khan directed us to the other room which was furnished with new *Lahaf va Toshak* (Persian quilt and Mattresses) round pillows and clean loose bed sheets.

We all were tired and as soon as we were under our *Lahaf* we were deep asleep, next morning when we were awakened by the sounds people around us were making when they brought *Manqal va Qoori Chahi* **[4] in the room for breakfast, fresh butter and wild honey, milk and *Noone Tiri*. Before we started

our breakfast the Monchy served each one (*Estekan*) (*a small glass tea cup*) of boiling water with pinch of sugar or honey in it, of course here again I had to ask Masoud what this was and why they served hot water, *he explained*: In the mountains where the air is very dry, in the morning before breakfast they serve boiled water instead of tea to clear your throat and if you have a little it could help you to ease down your coughing, and moreover you feel much more awake after having this. ***(I recommend the readers try this)*
We packed our gears again and said our farewell to Ali Mardoun Khan, it was clear sky, nice weather, with the temperature not exceeding 70 degr. F. Beautiful travelling weather. Fresh from a good nights rest and on the road again. After two hours we dismounted and were walking our mules when Cheragh Ali again started to sing:

"OVIDOM SARE KUH, SAIL KERDOM, LASHE KHODA BAS MENE OU DIYARE, "DAY BALAL LOM DAY BALAL, SOUZE TIYE KALUM DAY BALAL."

I remember clearly my feelings as we rounded the river bend into the village of Dashtak. This was the first village with a road which served motorized vehicles of any kind, which we had encountered in over three weeks. We had travelled almost 1000 kilometers across some of the wildest country on earth, had the adventure of a lifetime, and I wanted.... A pepsi! Here in Dashtak, were small businesses, bakeries, mosques, a public bath, a hotel, grubby little houses which looked more like hovels, and a RESTAURANT. Here we weren't allowed to display our weapons openly, and I wasn't the oddity which I had been for the last three weeks, Civilization.

We lunched at the little restaurant on skewers of kebab, plates of noon, piles of rice, vegetables both boiled and roasted, and Pepsi Cola. When you view Arabic and Persian scripts, the typical westerner, thinks they are the same. They are not. In thc Persian alphabet contains four letters which Arabic does not have them. There is a (P), (CH), (Z), and a (G), as near as can be translated. In Arabic lands there is no Pepsi, it is Bebsi, because Arabic has no P sound.

When we left Ahwaz, on the west of the high Zagros, the Persian language was spoken with an Arabic accent. Pepsi for instance was spoken with a strange admixture of P and B sounds, not clearly either. Here in Dashtak, it was on the eastern side of the high Zagros, a completely different cultural heritage, and Pepsi had a clear and ringingly distinct P sound.

Persians can be a seriously teasing people. This is often mistaken by travelers, because they don't realize that they are just having fun. It is too often seen as the local people having fun at the foreigner's expense!. Here the propensity to not worry about Taarof around foreigners comes into play. They will tease, until the foreigner invokes Taarof themselves, then the Persian will be ashamed to let his manners be shown lacking, and become very effusive.

I once took my family to the capital city of Iran, and we stopped in a local restaurant for dinner. It was obvious we were American, and the feraj, or waiter dutifully brought our menus out. Most restaurants in Iran had dual script menus. One side written in the Persian language and the other side in Western Script. They were having a bit of fun with the foreign family. I was one of the very few Americans in the country who not only spoke Farsi, the Persian language, but could read and write as well. It had been said I read like a third grader, which I took as a compliment.

When I ordered the meal for everyone, like a Persian would, they were amazed. During the dinner, they kept changing waiters. After about the fourth change, it was too much for me, and I asked what was going on. The waiter quickly apologized to me, and explained that in all the years he had been working there, he had met very few American who could speak Farsi, but, none of them had ever met an American who spoke Farsi with an Arabic accent.

After the surprise of hearing the clear P, the experience consolidated my understanding of what we had done. We had crossed one of the major mountain chains of the world, on foot and horseback, and arrived safe, healthy, and, after three weeks in the wilderness, looking very much like wild men ourselves. I think I felt like Jules Verne's famous character Fileas Fogg, who arrived home very disoriented after such a journey.

After we lunched, our mountain friends took their mules, horses and the small money we paid them as bonus for our safe delivery and returned up the road to their mountain homes. It was a sad goodbye indeed. During this time, I had not realized how fond I was becoming of some of these men. When I later learned of the accidental death of one of the men, I was in mourning for some time.

Our faithful Mahmeed had been off seeking transportation to take us to Shah-re-Kurd town where we could catch a bus.

After making our goodbyes, Mahmeed pulled up with three jeeps with our gear and baggage already piled in the back of one. The five of us piled in, and in a cloud of blue smoke we took off to meet our bus in Shah-re-Kurd.

In the early 1970's, , air service from Esfahan to Ahwaz and return was only on a weekly run. Our flight was due to leave the next morning to return us to our homes and families. It was planned that Mahmeed and Ahooie would either secure the services of a truck which they could ride in, or they would return with all our gear in a bus from Esfahan to Ahwaz. They, of course, could not take the weapons. The weapons had to stay with Masoud and I because of the very serious view of unlicensed weapons in Iran at that time.

Well, one of the jeeps had a breakdown, and Mahmeed got motion sickness on the twisting mountain road. The Iranians are quite superstitious people, and they have some strange beliefs. When Mahmeed got motion sickness, the local people advised him to drink a shot of gasoline. Masoud, Dr. Fathi and I all advised him against this, but he was un-swayed in his intentions, and drank about a shot glass full of gasoline. He obviously became really sick then, and we arrived in Shah-re-Kurd, too late to catch a bus on to Esfahan. That night we took a hotel room in hopes of securing transport early enough in the morning to catch the flight out of Esfahan.

Dr. Fathi and Masoud at the 15th century Abbasid Bazaar in Esfahan

It was a small dingy hotel that we found. No bathrooms, tiny little cots, which we put our sleeping bags on for a softer and cleaner sleeping surface, but they didn't have stones in the mattresses! For the first time in over three weeks we experienced a real bed. In the early morning we awoke, had breakfast, and without the help of our faithful Mahmeed, started to seek a bus. There was supposed to be a daily, early morning bus to Esfahan, but its' reliability was – (In Shah Allah), or God willing. I guess today God wasn't willing, because we didn't find a bus until it was too late to make the flight. It was probably as well because Mahmeed was still sick and Ahooie was "visiting" someone, and not found until in the afternoon.

We finally secured a bus to Esfahan, and arrived there about 3:00 in the afternoon.

After putting the gear under Lock and Key, making sure Mahmeed was good for the night, we went and bought bus tickets for the following morning. Masoud and Dr. Fathi then took me to the marvelous Esfahan Bazaar. The ancient bazaar in iran is something to behold. Because of the very hot temperatures, the bazaar, or market is built underground. It stretches for miles, and is full of shops containing anything and everything you would expect in any market in the world. It is truly a Mall of Malls.

We bought some cashew nuts, and pistachios, then found some drinks and GAZ. Gaz means "gas". It is a sweet made out of nougat, with pistachios whipped up and then cooked. It is truly as light as gas. We also found some "Chus-e-fil", and bought some for the evening. "Chus-e-fil", is popcorn. Its' name, however means literally "fart of elephant". It seems that when popcorn was imported into Iran in the late 1800's it was related that it was as light as the fart of an elephant. The name stuck.

We had a marvelous night in the hotel, and went to catch our bus the next day.

When traveling in Iran, you must always carry your documents with you. Your "Shenas-nameh'", documents passport, shot record, work permit, residence permit, gun license, etc. We needed all these things for me to buy a bus ticket. Not so much if I were going to travel by air, but busses were very regulated.

After boarding the bus, and stowing our considerable gear, we took our seats across the back of the bus. We claimed the entire back seat of the bus, because we were not the ordinary travelers. We didn't have any chickens, goats, or baskets of fruit, just had Guns, camping gears, and travelling stuff. We also didn't have any food. Nobody told us that we were supposed to bring our own food for the near 24 hour bus ride. There were two drivers, who took turns driving, and the only stops were for fuel, toilet, and driver change. When the bus stopped, we had to immediately run to a semipublic latrine, and go back to the bus in shifts to guard our possessions. We were able to buy cookies, bananas, (the only safe fruit to buy in public), and soft drinks, but the stops were hurried, and no restaurants were available. They surely didn't have McDonalds or Burger King along the road.

We were sitting in the back row of the bus, when a young man boarded at one of the stops. He brought a basket of goods to sit beside him in the seat. Masoud was joking with him, and said, "as soon as you go to sleep, I may take some of your fruit". The young man took this very seriously, and replied, while rolling up his very dirty sleeve, "here you may have some of my blood as well". This was a very serious reply, even an insult to a joking attempt to break the ice, and the whole bus turned cold. Nobody spoke for several hours, until this young man got off the bus at a fuel stop, and wasn't allowed to re-board. The bus left without him.

Finally, early in the morning five weary travelers returned to their home town. We got taxis to take us to our homes, and slept for a couple of days.

Chapter 18

After the Big Hunt

Masoud and I hunted wild boar, Sheep and Ibex on a nearly weekly basis. Masoud would pick me up from work at 2:00 in the afternoon, after my 8 hours of summer time work, and we would go a two hour drive to the mountains to hunt sheep. Masoud had a more flexible schedule than I, and he would have our guns, and gear in the car, along with two 20 liter water coolers with ice in them. We would zoom off to some low mountains nearby Ahwaz and hunt until dark. We always found sheep or ibex almost at dark. And we would take our game, and then walk back to the car, and drive home.

When we first arrived in the mountains, we would hire a coolie, or fellow to carry our stuff. We left one of the water coolers in the car, and the coolie would carry the other one on his back. After finding and shooting our sheep, we would clean them, then wash, and cook some liver over a small cooking fire. We ate supper there with some yoghurt, bread, onions, and of course roasted liver. We drank water, and the coolie would then carry the sheep back to the car for us. By the time we shot the sheep, and had supper, the water was nearly empty, so Masoud and I would take turns carrying the near empty cooler, and we would walk back to the car in the dark. When we arrived at the car, we had about 20 liters of cold water to drink on the way back to our homes. By the time we arrived at about 10:00 in the evening, all the water would be gone. It was a hot, dry country.

We had lots of evening and weekend safaris, but we desired for longer, deeper into the mountain hunting trips. The next summer, Masoud arranged for us to go to the high Zagros Mountains again, this time from the Esfahan side.

He wrote letters, and sent telegrams, even sending guarantee money to a man who was to guide us into the high mountains for a two week hunting trip. It was about a 6 hour drive from Esfahan to the village where we were to meet our guide. There was another village, which had electricity and running water, called Farsan along the way to our jumping off point. We had arranged to meet in Farsan, rest a little, and then go into the mountains.

We left Ahwaz on a Friday morning. Friday is the Sabbath in Iran, and most of the Middle East, or Muslim countries. For westerners, it is sometimes difficult to remember that a new day begins at sundown. The Jewish people make the time more exact, noting that the new day begins when you can see three stars in the sky at the same time. We needed to be at our destination because the next day was a very religious holiday, the day Ashura, when Shiite Muslims mourn for a massacre of their original Imam and his followers after Mohammed's death.

Before we left, Masoud made all the arrangements for travel. We had to have travel documents to go back into the mountains armed, because the government was very concerned about rebellion. He had already alerted his friend, the governor-general of the Esfahan province, that we would be bringing guns through his airport, and when to expect us. We expected to immediately board motor transport to drive about 4 hours where we would overnight. When we arrived in Esfahan, just before noon, we were met by officials from the governor-general's office.

We thought it was very nice, as we had expected to have some trouble getting the guns from cargo, and that meant we would have to give some "help" to the unfortunate people in charge of cargo. The arrival of higher level officials cleared the way for all the gear and we were cleared immediately. As soon as we had our trucks packed with all of our offloaded gear, we were set to go, but the governor's officials wouldn't hear any of it. We had to go have a "light lunch" with the governor. When the governor asks, you don't refuse, so off we went with 3 trucks, and we four "Shekar-chi" or hunters. Masoud and I were invited and accepted as honored guests, and Mahmeed and Ahooie were relegated to the working people to entertain. We passed a pleasant time with the governor, because as always, he loved Masoud, and was enthralled with a "ferangi" or foreigner who obviously loved the Iranian people. The lunch turned into afternoon, and it was left to me to break ***taarof***, and beg the governor and his staff for his leave for us to make the journey to ***Farsan*** village, so as to arrive well before dark, and the coming holiday.

By the time we got away, it was nearly 4:00 in the afternoon, and we thought had a good chance at making our destination before dark. Fate had yet to play its trump card. About an hour from the Farsan village, we encountered road damage. By the time we had worried our way through this damage, it was dark, and by the time we arrived in Farsan, it was well into the Ashura holiday.

We were going to meet ***Safar Torrabi*** one of the respectable Bakhtyari businessman trading between Bakhtyari Nomads and the city traders that mostly situated in ***Shah re Kurd,*** selling all kinds of articles, victuals and buying from Nomads milk, butter, wool, carpets, live stocks etc. most of these kinds of trading based on barter deals.

It was pitch dark when we arrived at the outskirt of the Farsan village, no electric lights in the streets and because the village generator had already been turned off, we had to use our flashlights to find our way, no one was outside in the narrow streets and we did not know the way to ***Torrabi's*** house, At last Mahmeed knocked on the door of a house we came across, after 10 minutes of continuous knocking loud and louder at last someone called from inside the house and asked us ***Kini*** (who are you) ***Chi khoy*** (what you want) Mahmeed with the Bakhtyari language told him that we just arrived here and we are looking for Safar Torrabi's house.

Then the man opened the door and he said he would come with us to show the way, we followed him and after passing through several mud brick houses at last he stopped in front of a large gate. After several tries to arouse someone by knocking on the door hard, someone from inside called and the man told him that he had Torabi's guests with him. Then the gate opened. Because I was the foreigner, and could be excused for making such a faux-pas, I told people that we were there to see Torrabi. We were escorted into the yard, parked our pickup trucks, and saw the pickup truck with our camping gear that Mr. Ahooie had arranged sent on road from M.I.S. via Khorramabad, Esfahan, Share Kurd to Farsan to meet us upon arrival.

Mr. Torrabi came out to greet us, he said that he was expecting us since the previous week and he had sent messages to the villages on your way to let my relations know that you were coming and they should take good care of you and show you the way to my place.

Then Mr. Torabi said I am sure that you are very tired, he ordered one of many rooms to be prepared for us, and ordered his Monshy to arrange all facilities of our comfort. As we spoke with Mr. Torrabi, we stressed to him that we wished to go to the Canyon of Qaisari, where we were to meet our guide to go into the mountains.

Mr. Torrabi strongly advised us not to go there at night because the road was very dangerous. Going through the narrow road with sharp cliffs, drops of more than two thousand feet would be very dangerous moreover there would be many heavy stones in the middle of the road to be cleared while driving. In the end, he became adamant, and said that even if the road was safe he would not let us travel at that time of night. After some more discussion we decided to stay for the night and indeed it was a gift to be able to sleep in a warm bed.

Mr. Torrabi advised us privately that the man we had contracted with as a guide didn't have a very good reputation, and that he was sure we would be back and stay with him before the week was over. He was so obviously concerned about our welfare that we realized he was not just exercising taarof, and we began to worry.

We were ushered to our community sleeping room, or guest room with many mattresses, and bed covers rolled up into their own place on the floor. As we prepared to go to sleep, a servant came in, telling us that he had been charged to make a chicken and rice dinner for us. They brought in a ***Manqal ,*** which is a rectangular tray on 4 feet about 2-3 inches deep filled with burning charcoal. It also had a teapot on the side, with the base banked in charcoal.

The tradition of tea or "Chai" is strong in Iran. I don't drink tea or coffee, because of my religious preference, and learned to either drink water, or soft drinks if offered. On this occasion, I drank a cup of sweetened hot water, and soon the chicken and rice arrived. The servants brought in a big tray with napkins, a big pile of rice, and broiled chicken. They unrolled a large plastic table cloth on the floor, and we sat down around that tray. After the obligatory "Be Essem Allah – Rahman E' Rahim", meaning in the name of God we begin, our party began to eat.

Masoud and Mahmeed were concerned because they knew that even though I was hungry, I wasn't eating. There were no utensils. The chicken was torn up, and there hadn't been time to bake the flat un-leavened bread that was normally used as a spoon to eat rice. I simply didn't know how to eat rice with my hands! I had not been initiated yet.

Eating rice with your fingers sounds like a simple thing. It isn't. You have to use your thumb and two first fingers and without bending any joints, pull the rice into a pyramid shaped pile about an inch and a half high. As it is thus compacted, the rice becomes slightly sticky. You then pick up the little pyramid, and placing your two fingers on your upper lip, you push the whole thing into your mouth and eat. It sounds simple doesn't it? Try it! I had rice scattered all about me, the other guests were laughing at my efforts. What a comical sight it must have been.

Normally the Persian Taarof wouldn't permit a person to laugh at another's distress in this way. I however was raised by very good parents. I remember, when I was embarrassed at an early age, my father taught me not to be afraid or embarrassed, but to laugh at my own mistake, or shortcoming. That way nobody laughs at you, but rather with you. I have found that advice to stand me in good stead all through my life, and turn stressful situations, into bonding moments, when laughter replaces anxiety.

After the rather messy dinner was over, we unrolled our sleeping mats, and blankets, and slept the sleep of babies. In the early morning, Mr. Torrabi came in with his servants to awaken us with hot water, "Gand" or crystalline sugar which is consumed with tea or hot water, and yoghurt, cucumbers, and hot, freshly baked bread.

After a breakfast, we proceeded to get packed and ready to go to Qaisari where we were to spend two weeks hunting. The small earthquake which had happened the day before, and had caused us so much trouble on the road to Farsan, had almost destroyed the road to Qaisari. It took us from about 7:00 until nearly noon to go the 20 kilometers between the two villages. Mr. Torrabi had been completely right about the condition of the road. We had taken some laborers in the truck with us, and put them to good use clearing the road.

You might think that we were some kind of prima-donna personalities, not clearing the road ourselves. We were all capable of hard work, and at first I was annoyed by my friend Masoud insisting that we hire people to do this work. Time after time, I have come to find this man as the kindest man I have ever met. He looked at these challenges as an opportunity to give a job and some money to people of far inferior means, who would not otherwise have any cash money. As they lived their more or less primitive existence, they bartered for everything. Clothing, shoes, tools etc. Cash was almost unknown in the mountains, and they were very grateful for the opportunity to earn a little.

When we arrived at Qaisari, Mr. Torrabi's other predictions became more and more obvious. The painstaking preparations that Masoud had gone through to have mules, and a guide ready to take us into the mountains were for naught. There was not a single mule in the village, and the guide was not to be found. "The Colonel" suddenly became clear to view. I have seldom seen Masoud really angry. This time he was really angry. The village was in a sudden activity to find the offender. He was quickly brought to us for an audience which he had thought to avoid. In front of the entire village, he was told that he had to get the mules ready by nightfall, and we would leave before daylight in the morning. He complained that he had already spent the money, and he didn't have any funds. He was told to get the mules, or go to jail. At about 4:30 in the morning, a very tired and shaken man came into our camp to present himself along with the 15 mules and 5 helpers he had secured for the trip.

We proceeded to load the mules, and by 5:00 am, we were heading up "Tang Qaisari" or the canyon Qaisari. We traveled hard for 3 hours and stopped for breakfast. It was a very quiet trip. None of the singing and banter which normally accompanied a safari. Everyone was very much on edge. We breakfasted on cheese, yoghurt, water and bread, then on the trail again. It was obvious that something was wrong from the demeanor of the group. We travelled until about noon, then shot some partridges. Although we stopped every 15 minutes and looked for sheep and Ibex, we saw nothing but a few partridges and gazelles.

Masoud and Mahmeed went away, and sat in a private conversation for about a half hour. And while they were gone, the mule handlers prepared the partridges for travel. To keep them from going sour in the heat of the day, the fellows put a thorn up their anal cavity, and pulled the entire gut out without disturbing anything else. Without this treatment, before evening, they would have become bloated and spoiled. This way, by the time we got to the evening camp, we could cook them with rice, bread, and some vegetables.

Masoud and Mahmeed returned from their conference, and it was clear that something momentous was up. A discussion about where we were going quickly turned into a heated argument, and I made sure I was very near my shotgun, without actually touching it. I have never seen Masoud and Mahmeed so angry.

The discussion and argument was about where we were being led. It was quite obvious that there was no game where we were, and Masoud and Mahmeed had discovered that we were being led away from where the real game was located. The argument seemed to be going nowhere, and everyone was becoming more and more agitated. I finally interrupted and asked Masoud, "why do we need this man, he has provided the mules and the handlers", Mahmeed is a better hunter and guide, even though he doesn't know the country". That was all it took to cool the discussion. The guide was dismissed and sent away.

I remember Masoud saying "We are going to go up on Qaisari mountain, and with these powerful rifles, take every piece of game off its' face". He went on, "we will kill all the game, and give it to the poor people of this valley". By this time I had my shotgun sitting across my lap, and they had all seen me shoot partridges. The steam left the old guide's boiler, and he slunk away, never to be seen again, with some stern warnings about causing us more trouble, and the inhospitable atmosphere awaiting him at ***Shah re Kurd*** prison, should we ever hear his name again, or if he caused any trouble for us. Mahmeed and Ahooie checked with every single one of the handlers, who supplied the mules, and made sure they had already been paid, and we changed our direction and went on.

We travelled until about 9:00 that evening. As soon as the group had recognized that new leadership was in place, they warmed up immediately. It was a very hard day's travel, and everyone was very tired by the time we reached our destination. We had travelled for the last hour and a half by moonlight, carefully picking our way through the stony path, to avoid an accident. We arrived a very tired group, although in high spirits. The handlers saw to their mules and sleeping arrangements, Mahmeed and Ahooie saw to our camp, and Masoud and I saw to supper. We roasted the partridges, cooked rice and vegetables, and made some cold dugh from yoghurt and water. By 10:00 pm. We had eaten a nice supper, and were bedded down ready to sleep. It is amazing how the change of reporting structure had cleared the air, and resolved the problems. It took Masoud almost two hours to get to sleep that night. He was a very angry man, but only his close friends could actually tell. His actions were always gracious, and directed to our goals.

Mahmeed woke us in the morning. It was cold, near to freezing, and hunting was in the air! We had Citizens Band Radios with us, and after we had seen the hunting area in the dawn's light, we quickly made plans to split up. We were camped on a huge escarpment, with vertical drop of about 300 meters, ending in a sloping plateau, which ended at another escarpment of several thousand meters. The plateau was about a kilometer long by 300 meters deep. We decided to have Masoud and Mahmeed go to the east side of the plateau, and Ahooie and I go to the west side.

There were two massive Ibex with horns over a meter long facing off and bleating at each other on the plateau. They were clearly preparing a battle royal for some goats were there waiting proceedings. The Ibex cordially awaited our getting in position, but just about the time Masoud and I were both in position, they began to fight. They would back off about 20 meters from each other, then in a prancing dance, go around in a circle, finally ending in a horns down rush to meet in the center of the circle with a crash. The resounding crash could be heard echoing up and down the escarpment.

The bigger Ibex, soon bested the smaller one, and we chose which Ibex was the best target for our present elevated position. I had been calculating the distance, and the angle which we must attempt our shots. We communicated the trajectory of the 500 and 600 Meter shots, and decided we would shoot on Mahmeed's count of three over the radio.

Mahmeed was not the king of technology. When using a push-to-talk radio, he was constantly have to be told to push the button to talk, and wait until the other person was finished before trying to talk himself. How many times had I heard Masoud remind him "Mahmeed, harf bezand", or hit the button. We would fall into fits of laughter to hear Mahmeed tell us to listen by saying "be gusham", instead of "be gusam". His Farsi was overcome by his Backhtyiari language, and he was likely to say "fahrt" rather than "listen". What a hoot this wonderful man gave us. Mahmeed had the keenest vision of any man I have every met. He could see game unaided where none was apparent to a normal man scanning with good binoculars.

On this occasion, Mahmeed indeed held the button down, and counted to three – (Yek; Do; Se). on three we both fired and both Ibex went down. One fell, then immediately got back up and started running. Masoud fired again, and it fell, then started to slide towards the lowest escarpment. We thought it would finally stop, but it just slid over the edge.

Our group of handlers were jubilant. We had promised them meat, and they saw meat. They quickly went down and gathered the Ibex which stopped on the plateau. Three of them went over the edge of the bottom escarpment to retrieve the one which fell to the valley floor. After some time they came back to shout to us that it had miraculously fallen to a ledge, then taken off running again. They chased that Ibex into the night, but never retrieved it. The only thing we could think is that it was hit in the horns, knocked unconscious and then the shock of the fall brought it around again.

We hunted the rest of that day, accompanied by Mahmeed, Ahooie, and the two handlers. We took several other Ibex, and spent another cold night on that high mountain. In the morning, we again took stock of Mr. Toorabi's warning, and decided to go to the village, gather our things and go back to Farsan.

We took our 5 Ibex, and returned to Qaisari to pack up and go. We took up a camp in the village, inside a grape arbor, and cooked some meat, then sent Ahooie to the neighboring market to get some things to go with lunch, and arrange for some transport to take us back to Farsan.

Mr. Ahooie was always keen to impress. While in the market, he called back to Masoud on the walkie talkie. These Citizens Band Radios had amazing range from mountain to mountain in Iran, because they had no interference.

Mr. Ahooie was only about 5 Kilometers away from us, and he was having fun with the crowd around him, who had never seen such a thing. Not above a little exaggeration, he told them he was talking to America, because he was speaking in English, they believed him. One of them ran and fetched a gendarme. The next thing we knew Ahooie's startled voice came to us, saying he was caught, and coming to us.

In about 20 minutes he arrived, in the company of a gendarme, in a rented jeep station wagon. We quickly explained to the gendarme that Ahooie was exaggerating for the benefit of the local people, and that the radios were perfectly legal and above board. This particular gendarme was a bit denser than most, and decided to make a big deal about it. He arrested the lot of us. After he made the arrest, he realized that he had arrested a foreigner, and things could get a bit sticky. When he found out about the "Colonel", they were real sticky.

He decided to transport us to the army base at ***Shah re Kurd***, and told the driver of the jeep station wagon to take us there. I had one of the walkie talkies in my hand, and he had the other one. He was afraid to ask me for the one I had, and I didn't volunteer. I just got in the back of the jeep and acted stupid. We were transported 4 hours to ***Shah re Kurd*** army base, which was the central area command for the region.

As we travelled, I opened the radio I had, and turned four of the nine batteries in reverse, leaving only about 1 to 2 volts to power the radio. We arrived about 5:00 in the afternoon, and all the brass had gone home. He put Ahooie in jail, but didn't know what to do with Masoud and I. one of the junior officers told him he was in some trouble, and suggested that for his own benefit he put us on our own recognizance. At that point, he took the radio I carried for evidence, and we were escorted to a nice little hotel, and told he would come back and pick us up in the morning at 8:00 am.

Masoud and I were glad for a shower and a nice evening, so we made the best of it. We cleaned up, and changed clothes for dinner, then went downstairs. When we arrived downstairs, in the dining room, we sat down, ordered soft drinks, and looked around. I noticed a military officer wearing the insignia of a full colonel, and asked Masoud what the rank of the base commander would be. He verified that he would be a full colonel, and so I just acted stupid, and brash, and went over to the officer. I mentioned to him that I recognized his rank, that it looked like US military insignia, and the conversation was on! He was there on single status, lived in the hotel, and was bored to tears because of his rank.

I called Masoud over, and we talked about his training of two years in the US military in Georgia. His name was Hormuz, but he told me that everyone in the US couldn't pronounce it properly, so they just called him Hank. We had a very pleasant evening, until about 11:30, when we paid for dinner, and went to bed. When Hank asked what brought us to ***Shah re Kurd***, we just said it was some administrative problem over licensing, and we came to clear it up.

When the gendarme came to pick us up in the morning, we were at breakfast, and made sure we delayed enough to let Hank get ensconced in his office. At 8:30 we were taken to Hank's office for trial and disposition. Imagine his surprise when we walked into his office, under arrest for international espionage! He looked at us, and winked, so we didn't let on we knew him at all. The gendarme presented his case, which was that he had witnessed Ahooie speaking with America on the radio and did his duty. Colonel Hank took the pair of CB Radios, and looked them over. He explained to the gendarme, that these held a public license, and were good for only a few kilometers, and that they were useful, but basically "children's toys". The gendarme was adamant, and so Hank said that he would make a test. He took one radio, and told his aide to take the other one and see how far they would communicate. They worked fine in the same room, but as soon as the aide got on the other side of the wall with the disabled radio, it was a no-go.

Hank brought the aide back in, and proceeded to give the gendarme a dressing down for inconveniencing these two obvious gentlemen. Oh how the mighty have fallen. These rural gendarmes play havoc with their community, extracting bribes, and generally making their life difficult, just for the fun of it. The gendarmes are not paid much, normally enough to pay for their cigarettes. Everything else is supplied for them, and they don't have to pay for much in their community. The gendarme complained that he needed to have money to pay for about six months salary on the jeep rental. The colonel told him to find it when he got back to his village and dismissed everyone.

When we arrived back in Qaisari, everything was ready to go, and Masoud paid for the jeep rental. His philosophy was that we might travel this way again, and he would rather have a gendarme owing him, than hating him.

Off we went to Farsan, where we arrived in time for a nice meal at the Torrabi home, and a pleasant evening. We explained to Mr. Torrabi what had happened, and everyone had a good laugh. We were shown around his shops and the village, then Mr. Torrabi brought his personal friend to us for discussion of the morning hunt.

Farsan was much lower elevation than Qasari, and so there were no Ibex there, only mountain sheep. We left early in the next morning, and hunted for four days, and returned to Farsan on a Saturday. We brought with us five freshly killed mountain sheep, and the remains of one which we had been using for camp meat. These were big sheep, at least twice to three times the size of a tame sheep, and highly desired for their wild qualities. When we arrived back in Farsan, Mr. Torrabi was presented with the whole thing. He immediately sent word out to the whole village of about 2000 people to come and get their share of the meat.

Mountain sheep near Farsan Village

Mr. Torrabi was like a Khan, or headman of the village. I suspect we would translate his title as Mayor. Every household in the village presented themselves to him in order of their importance, from the gendarme, to the widows. Each was hacked off a portion of the wild sheep, and told to tell the foreign guest and his friend their appreciation. It was all very touching. When we finished, there was about a half a sheep left over and Mr. Torrabi claimed that.

Farsan was a progressive village up in the mountains. It had a generator. At about 6:00 pm, the generator was started, and each home had a single electric light bulb for the occupants to prepare dinner, study, read, and let a little refrigerator, if they had one, cool food and drink, and make a little ice for the next day. At 9:00 the generator was shut down for another day.

Hunting in the mountains above Mr. Torrabi's village.

We stayed with Mr. Torrabi for several days, making day trips sightseeing around Farsan, and little hunting forays close to the village. Each evening we would come back into the village and experience a treat which I had never dreamed of; a Persian public bath house. This bath house was built in sections for different socio-economic classes. Hot water came from a spring, and you could go in a bask in a swimming size bath tub. What a luxury.

During the time we were staying with Mr. Torrabi, we took a lot of photographs. He was tremendously interested in my expensive camera equipment. I also had a polaroid camera, and showed him and his family several snapshots which I had taken. I had portrait pictures of my wife and four daughters and those caused lots of consternation. In public the Muslim household is very circumspect. In private, they are much like everyone else. After a few days, Mr. Torrabi confided in me that he would like some portrait photographs of his wives and children. He set guards about the place, and one afternoon, when the light was cooperative, I took photographs of his entire family. When we returned to our home, I took the film into the darkroom, developed it, and printed some very nice 11x14 portraiture for him. I sent the photos by courier service, so they would not be subject to casual post office employees, or damage, each in a nice frame. He was grateful to the extreme.

***Mr. Torabi's Monshi (Secretary) in his trade center.
The local Bakhtyaris would bring in agriculture goods
to trade for his shop contents.***

This time when we returned to Esfahan, we caught our plane, and Mahmeed and Ahooie drove the pickup truck back to Ahwaz in a leisurely manner.

Chapter 19

Sharokhi's village

In the autumn of 1976, two years before the beginning of the Iranian revolution, Masoud told me that Ruud Paauwe, Manager of the Schreiner Helicopter Services had had agreed to give us a lift (a two way run over the Zagros Mountains) to drop us near the village of Sharokhi, in a big helicopter, and pick us up ten days later on an exact day and time, to bring us back to Ahwaz.

For some time Ruud had lived in Tehran, in the north of Iran. As he travelled in and out of Iran on a regular basis, having a firearm was a big problem. Every foreigner who owned a licensed firearm in Iran faced the same problem. When we brought that firearm into the country, it was entered in our passport, with the entry; "Must take this firearm out of the country on final departure". If you travelled a lot, that really meant every departure, because the poor passport control officer didn't know if it were your final departure or not.

The answer to this was to store the firearm with the secret police during your trip abroad, and then retrieve it upon your return.

Ruud didn't know this, and on his first business trip, he showed up at the airport, with his shotgun locked up at home. Passport control turned him away. He quickly ran out of the airport, got a taxi and travelled about a half hour to his house. He took his shotgun in it's canvas bag, and ran back to the airport. This time they checked that he had the requisite shotgun, and because he was late they quickly directed him to security; carrying the shotgun in its' cloth bag.

Security was very thorough. I have about a hundred pictures of smiling security agents, because you either had to show them that it was a real camera you were carrying by taking a picture, or they take the camera apart. If you didn't want another picture of a smiling security agent, you had to make sure your camera was empty before going to the airport.

I don't think they were extremely flexible in their training, because when Mr. Powwe showed up, they quickly frisked him while he was holding the shotgun in his hand by the barrel; inside the cloth bag. Anyone remotely familiar with firearms would have recognized it for what it was.

When Ruud got on the KLM flight destined for Amsterdam, which they were holding for his late arrival, he was quickly ushered to his seat, still holding the shotgun. There wasn't room for it in the overhead bin, so he held it in his hand, butt stock on the floor of the plane, with the muzzle pointing up, in the aisle next to his seat.

After takeoff, he realized he didn't want to hold this gun all the way for a 7 hour flight to Amsterdam; he tried to get a steward to put it in the closet. They kept running up and down the aisle, serving passengers, with a curt, we don't have time right now, after a minute please.

When they finally got back to him, he explained that he would like to put his "package" in the closet, so he didn't have to hold it. They asked him what it contained, and he told them. There was a minor panic in the cabin. They took the gun and stored it with the pilot. Rude reports that it wasn't a boring flight for him, he spent the whole flight filling out forms about how he got the gun on the airplane.

Ruud was a good friend of everyone, and a hunter himself. He only had a shotgun, but hunted waterfowl, partridge and pigs whenever he could. Accompanying us was an English Co-Pilot, and another friend, Arjang Safiri, who would return to Ahwaz with the helicopter after dropping us off.

An early helicopter used to traverse the high elevations of Iran. No extra weight here!

After a 55 minute helicopter ride, we landed in a clearing near Sharokhi's village. There waiting for us was the faithful Mahmeed. He had driven the long way around in a pickup truck and had all of our gear, except for our rifles and ammunition. I had been very busy with some business, and Mahmeed had been busy with his job, so we hadn't seen each other for about a month. What a happy reunion it was to see him.

After Rude took the helicopter away, we were driven directly to the village, where we were welcomed by the village elders. This was a different tribe of Bakhtyari than we had stayed with before, and their language was quite different. Even Masoud had some trouble with the language. Mahmeed, however had no trouble at all, and served as interpreter when we were unable to establish communication.

We were welcomed like visiting royalty, into the general meeting tent. This tent served as a local government assembly, and consisted of a huge log, which was originally about 50 or 69 centimeters in diameter, (20 to 24 inches). This log had been cut in half lengthwise, and formed a very long single seating area which made a V shape on each side of a central chair. That way, the village headman, who sat in the very center, could be seen by everyone as he conducted business. None of this seating had backs to lean on, and I never saw any Bakhtyari seated with a backrest, except when they sat on a floor with their back against a wall.

We were invited to sit with me on the Headman's right and Masoud on the left side in honored positions. This is where the language difficulties asserted themselves. I was struggling so hard to understand the barrage of questions put to me, that I forgot my Bahktyari manners. Desert climates have so little water, that bathing is nearly impossible. Even simple washing hands becomes difficult. With the advent of Islam 1300 years ago, an entirely new custom was introduced. Before each of the 6 prayer times, the devout Muslim was to do a ritual washing. In the some areas, this was more difficult than others. This lack of washing water, and hygiene in general went all the way to the community food bowl.

I had been taught many of the tenants of good manners in the Muslim world. I knew that in some families, or small communities, a community rice bowl, with pieces of bread, vegetables, even stew were shared. Since the Muslim community doesn't generally follow the western habit of cleaning themselves after using the toilet with paper, they use their left hand; either with or without water. The left hand is therefore considered to be foul, and not to be used in a way that westerners consider normal. I sat amazed, on many occasions, watching food being served, prepared, and eaten in the villages, all without the use of the left hand. Consider what happens then when a thief is judged and his Right Hand is removed by the stroke of a sword. He is then banished from the community food bowl, and condemned to a life of family or community isolation.

Lunch was being prepared, and a young boy brought the hand-washing materials to us. As guest of honor, I was expected to go first. Hand washing materials consisted of a large bowl with a filigree grate over the center, and a long necked pitcher of water. There was a small piece of soap sitting on the apron of the filigreed reservoir. One boy held the grated top reservoir in front of me, and the other proffered the pitcher. I was to take the soap in my right hand, then receive a little bit of water on that hand and lather up "the right hand only". The right thumb was used to wash the fingers, and then when finished, I was supposed to put the soap back on the apron of the grated reservoir, and receive another splash of water to rinse off the soap. Finally a towel was given, and it was swirled around the right hand with a flourish to dry the right hand.

When this ritual was proffered to me, I was struggling so hard to understand what was being said around me, that I forgot everything I knew about Bakhtyari manners, and just grabbed that soap, and lathered up both hands. I wasn't even looking at the boy holding the used water reservoir, until I saw the look on the village Headman's face. Of a sudden the realization of the faux-pas that I had just committed came rushing over me. I am sure that my face and neck turned several darker shades of red, but what could I do then. I simply held my hands over the filigreed dish, and received a rinse. Then I took the towel and dried my hands. Imagine the concerned looks of that row of village elders as I committed this sacrilege!

The Headman had been elected for good reason. He quickly assessed the situation, and without hesitation, he took the soap from the amazed young man, and lathered up both hands! He then took the rinse and elaborately washed and dried both hands. From him to the rest of the group went the washing materials, and every one of that group washed both hands. There were hands washed that day, which hadn't seen soap and water for many years. The water became so fouled and smelly, that the little boys changed the pitcher and waste several times before we were finished.

What I had just experienced was Taarof in its highest practice. In the Iranian custom they were "protecting the water of my face", or as we would say "saving my face". I witnessed an amazing set of manners or Taarof on many occasions. On other occasions, I witnessed some of the most abject cruelty, I ever will see. Let's just say the culture is very, very different than my own.

We were served a nice chicken lunch, and I was careful to eat in the traditional manner, so as not to cause any more problems. Then we brought Mahmeed into the center of our discussion and started planning the hunting trip.

All of the village hunters were present, and they all presented their ideas of where we should go. This presentation of ideas was not haphazard as it might be expected, but in a kind of pecking order, from youngest and least respected to most respected. All these hunters had muzzle-loader rifles, except a cousin of Colonel Sharokhi named Mansoor. He had an 8mm mauser caliber Mannlicher rifle, with iron sights, which was quite exquisite to see in this setting. He carried a bandolier of ammunition, unseen in the mountains, and it was his badge of honor.

Mansoor's suggestions were accepted and we planned to go out in the "baluch" forest the next day for Sheep and Ibex. These mountains and surroundings were very different that anywhere I had hunted before. The word baluch means oak, and the acorn is called balut. The province we were in was called "Baluchistan". Baluch for oak, and "stan" for state. This forested land was very different hunting country than the more open country with brush and small trees of the northern Zagros.

The rest of the day was to be spent in rest and preparation for the coming hunt. We slept in the afternoon sun, then as the evening got cold, we went into a tea-house to be with our friends. Outside it was cold, and inside it was hot. They had a wood fire going in a stove, and the room was stifling. As we walked in, I heard Masoud and Mahmeed say together "buh pah", which is stinky feet! It was indeed stinky feet. If I could change these people, I would have introduced a habit of a daily bath. This far from a public bath house, many had probably not been bathed since birth.

Not only stinky feet, but there were a large number smoking opium pipes. This was indeed a backward village! Masoud and Mahmeed were engaged in conversation with Mansoor our guide, and I took my rifle to the roof to get out of the press of people.

I originally had a high velocity small caliber 25-06 hand made rifle which I had carried. We had done so much long range shooting, and I had seen a good friend, Hadad, nearly killed because I couldn't cleanly take out a large brown bear, that I changed my rifle on the summer vacation in the USA. I returned with a 300 Winchester Magnum, again on a Mauser action. This was an extremely powerful and long range weapon. Masoud said it killed on one end, and wounded badly on the other end. It is quite heavy, and still with heavy hand loaded ammunition it "kicks like a mule".

When I brought this in through customs, with all the permits, letters and paraphernalia associated with my weapons licenses, we were met with a rush of customs agents. The normal Iranian people see so few weapons, that even the customs agents are enamoured. Every single one had to come and inspect the rifle, to hold it up to their shoulder, and say boom, boom.

Iranian customs are a sight. The agents are very poorly paid and are looking for any opportunity they can to extract a bribe to help the weary traveler through the ordeal which is called customs clearance. I have generally found this to be true in every country in the Middle East except Israel. The worst is North Africa, where the "dast" is openly solicited. Iran falls about half was down the scale. When my family, and several families whom I knew had all their baggage together, I told them to be ready, and I then took the gun case to the desk for clearance. When the rush of customs agents clustered around the clearance desk, my family and several others which had readied themselves, simply walked out of the customs hall un-impeded. What a hoot. It happened every time I took a weapon into the country.

When up on the flat roof, I just sat down against the parapet and began to look around at the countryside. I quickly observed a herd of tame sheep. They were being herded by 4 sheep dogs. I sat and watched the precision of these sheep dogs as they herded the sheep across the face of a mountain, towards a canyon. As I watched I thought that I had never seen a sheep dog in Iran. The Muslim people generally abhor dogs, and do not keep them as pets. They are kept as watch dogs, but they almost never establish the close relationship required to train them. As soon as I realized this, I took a closer look. They were wolves; big wolves. They were still a long way from the village, so I just grabbed my rifle and started shooting. I fired four shots, downed three of the wolves and one ran away.

Immediately the people began boiling up out of the building onto the roof to see what the shooting was all about. I had never seen a wolf in the Bakhtyari country, and had no idea what it's name was in the local language. I just shouted Wolf, which they took as woof. It took some minutes for Masoud and Mahmeed to instruct me that particular animal was called a "Gorg". As soon as the word gorg was pronounced the men put up a howl and ran to their homes. This high up in the mountains with a permanent village, which didn't move in the winter time to lower elevations, wolves are a real problem. They take sheep, goats, chickens etc. right out of the village. There are very few homes which have not experienced the loss of a child to the gorg in the winter.

Masoud was excited that I had shot the wolves. He asked exactly where they were. When I pointed out their location, the villagers became quiet. They simply didn't believe a wolf could be taken at a range of more than 500 meters. I am not sure what they thought but they definitely didn't think there were dead wolves on the mountainside. I think my previous misstep at the lunch had made them think of me as a bumbling fool. Our faithful Mahmeed, however, never doubted, but immediately told me to gather up my empty cartridges for later reloading, and suggested we go look at the dead wolves.

We filed out of that building like a group of angry bees. All noise and misdirection, with Mahmeed at the head of the line. Up the mountain we went. Before we even got there, we noted the smell. What a stink these dead wolves made. We found them there in the grass; three dead wolves, in the proximity of the herd of sheep which was milling around, not knowing what they should do.

When the villagers found that we could kill wolves they shouted with joy. All my sins were forgiven me. Quickly a wolf hunt was organized. We had a pickup truck and mansoor had a jeep. Masoud and Mahmeed rode in the jeep with Mansoor, and I rode in the pickup truck with our driver. The back of the pickup truck was full of shouting villagers. I would have liked to have a motion picture of this whole goings on. I guess the memory will have to suffice.

Down the road we went. By this time it was getting quite dark. Mahmeed was shining a spotlight around, and suddenly the jeep stopped. I told the driver to stop well back from the jeep, so our mob of villagers didn't get in front of a rifle and get killed. Masoud jumped out and fired two quick shots. I could tell from the sound that both were hits, but I didn't know for sure what he had shot. He could have been shooting at wild boar, which were plentiful, and loved eating the acorns. True to the evening, Masoud took two more wolves, and the celebration was on!

We went back to the village as conquering heroes. Not since alexander had such a celebration been seen.

When we got back to where we had left the dead wolves, only blood, and a few traces of bone were to be found. Every other scrap of those animals were taken by the villagers to put in their homes to ward off the evil wolves from their flocks and families. When we noticed this, we went to the back of the pickup truck to find the same thing in process for the wolves Masoud had killed. Masoud was only able to retrieve two canine teeth as a memento of that evening.

By this time, we were all tired, and retired to our sleeping bags. It did cause us to think about the guard around our camp. During this entire trip, we had no more problem with wolves. I also never detected a single incidence of anything but respect for any one of us. We all knew we were guests of a very powerful Khan, but we were accepted as comrades.

High sheep camp. Not much grazing here!

The next morning dawned crisp and clear. It was cold, and we awoke to our morning drink of hot water and sugar. Because this was a short trip, I had brought with me my morning drink; chocolate. It didn't last long. I am surprised how quickly chocolate is accepted by people who have never heard of it, or tasted it. As soon as they smelled the hot chocolate, they asked the Mulah, or Muslim Priest to rule whether it was clean or not clean. After he tasted it, and we explained how it was made, he quickly ruled it was clean and had his first cup. The can dwindled away quickly, and Masoud, Mahmeed and I were quickly reduced to drinking hot water and sugar as before.

The pancakes and syrup we were cooking for our breakfast soon met with similar fates. This whole breakfast scene, coupled with the wolf killing the night before, quickly put the lie to the old statement; "you only get one chance to make a first impression". We were much poorer in supplies, and I thought it would be ten days before I again tasted pancakes and eggs, and my beloved hot chocolate, but that angel Mahmeed had foreseen the debacle, and quietly squirreled away a small supply, which came out of hiding each morning to awaken me with a cheery greeting – Sob Bokeer, agha Taylor, or "good morning Mr. Taylor".

Masoud and Blaine in the mountains above the newly planted oak forests.

Don't get my meaning wrong. We wouldn't have denied these wonderful Bakhtyari people anything of our supplies. They would have literally shared their last food with us. That is the nature of the mountain people of Iran. They were simple but deeply good people with a culture which was very strange to me. As I remember my time with them, I love them, but despise their present government, which denies them the very basic respect due to people of any origin.

Off to the hunt we went. We went by jeep and pickup truck for about two hours, and then took horses and mules, with donkeys carrying our gear. We ascended into the higher mountains where there was more game. It took us until late in the afternoon to get into the hunting grounds. All this time we were traveling oak forests which had been planted only a few years before. There was a vast re-forestation program going on in this part of Iran. It was one of the Shah's great programs designed to return the country to, at least, a shadow of its former greatness.

When King Darius ruled in 500 BC, Iran had a dominion over all of the Middle East, and even some of Asia. One of the golden plates found in Persepolis recently starts: "I am Darius – King of Kings, ruler over 25 Nations". When Alexander the great came through in 323 BC, he broke up that empire. The inter-faith rivalry between Shiite and Sunni Muslim has done for Iran and the surrounding area what war couldn't do; divided and conquered the people.

We camped that evening on the heights above the oak forests. From this elevation up, things looked more like they did in the northern Zagros. Because we were a day's travel from the nearest permanent village, we were also seeing more game. We took a small wild goat to use for our supper, and cooked it with rice and vegetables. Everyone ate well, and retired.
When we awoke in the morning, our water supply was frozen over, and everyone was grumbling except me. I really enjoyed the cold weather. The other people were more used to the hot weather of Iran, and it was too cold for them. I dug out my heavy coat, breakfasted and was ready to go hunting.

Masoud and Mahmeed had been out at the crack of dawn, looking for a place to find the game. We packed up the camp, started the camp boys on their way to our next camp, and took up our weapons to go hunting. We took several helpers with us, to carry and run messages etc., and we started off up the mountain on foot. The sheep and ibex all live above timberline, in the rocks. They can't be hunted from below. It doesn't work. You must quickly attain altitude above them and hunt down on them. They almost never look up. Their habit is to find a peak or promontory, where they can lay down, and watch. Anything which is below them, they will see in plenty of time to clear themselves out of the way.

Getting ready to make camp and starting Dinner

One of the camp patrons making bread for dinner

Masoud on the right at a morning camp scene.

About noon, we found ourselves high above timberline, and started to see game. From this vantage point Masoud and I could take game from long ranges, with our modern rifles and telescope sights. The local hunters with their muzzle-loaders and even more modern cartridge rifles with iron sights didn't have a chance against these sharp eyed quarry. Unless you were a very capable outdoorsman and tracker, you couldn't get within 400 meters of one of these animals. We took several smaller goats, but didn't see any large sheep or ibex. Before dark we went back down the mountain to our waiting camp, where we cooked up the fresh meat we brought with us, and mixed it with the rice and vegetables already waiting for us. There is nothing like the high mountain air, with the smell of barbequed meat and a simple meal to make you comfortable.

A mountain stream near Mr. Sharokhi's village. Sharokhi and Blaine getting a drink.

The next three days of hunting yielded similar results. No big game animals. We decided that we needed to change our strategy. That afternoon we packed up and went back to where we had left the vehicles at the Oak forest. It was pleasant to stay that night at the lower elevation, where we could wake up to liquid water, and had plenty of dead wood to burn.

Our change of plans proved fortuitous, as we were nearing the village, we met a vehicle from town with a message. We were to return because the helicopter had to change plans and return early.

We arrived back in Sharokhi's village just in time to get our final packing done, and clean up a bit. We said our goodbyes, and waited at the clearing about 15 minutes, then here came the helicopter and we left for home.

Chapter 20

Home Is Where The Heart Is

Sharokhi's village was the last big hunt we were able to take together. There were many day hunts and weekend trips, but the climate was changing in Iran. There was a definite feeling that foreigners were increasingly unwelcome.

The people I worked with in the Oil Company were not aware of my command of Farsi. I found it better that way. In spring 1977, the orders came down from the upper level management of the Oil company, that all reports and analysis were to be submitted in Farsi, not the usual English. This brought the work to a virtual standstill. I notified the company that when I fulfilled my last year's contract in July, I would not be returning after the summer. I had been working on a big project, performing a petrophysical analysis of all the sandstone productive reservoirs in southern Iran. When I finished that work in March, I began the work of translating it into Farsi.

All of the Oil Company Persian staff were very afraid of doing any of these translations. It was well known that they would lose their jobs, if they were caught helping the foreigners. I did the translation myself, reverting to the shorter dictionary constantly, and working without help, because I didn't want anyone to get into trouble. I finished the translation in mid June, and handed the report in to the management of the technical group.

This caused quite a stir. As I was listening, I could tell that some very secretive detective work was going on. First they questioned all the staff, then they took all the typewriters away for analysis to find which one had typed the work. When they finally realized that I had typed the report by myself, everyone started thinking about what had happened in the past. Because it was assumed that the foreigners were unable to understand the language, the staff had become lazy, and would speak about confidential things in the presence of foreigners. It was very revealing, listening to the conversations when people thought you couldn't understand. The office staff were very divided, with almost nobody in the middle. Either they liked the foreigners, or they hated the foreigners. The few who really disliked us, generally did so out of jealousy. They seemed convinced that all foreigners got their good jobs because they had bought them or influenced others improperly.

In my 5 ½ years in Iran, I grew close to many Iranians, and many foreigners. I was president of the softball association, which was a big draw for the foreign community. At one time we had 57 mens teams, womens teams, youth teams, one cricket club team, and an Iranian team which I coached. I was also elected to be PTA president of the international school, and even played Santa Claus to deliver gifts to the international school students. My sleigh and reindeer was an Oil Company helicopter.

Masoud and I put together a PTA fund raising dinner, we supplied wild game meat for 5000 people and charged $50 a plate. It was open to the entire community. It was a great success, and provided money for a new school gymnasium, and all the sports equipment to fill it.

In all the time we lived there we only knew a half dozen American employees or contractors who had less than the highest integrity. When the foreign community left in 1978-1979, the Oil Business suffered tremendously. The Iranian staff were simply overwhelmed. Those jealous people who thought the foreigners didn't know anything were put into the vacated jobs, and a debacle ensued. The revolution brought about financial chaos.

Blaine with fish and game officers after taking an Ibex for the PTA dinner.

Chapter 21

The Relationship After Iran

Since leaving Iran in July 1977, Masoud and I have stayed best friends. I can only report that this man is the kindest, most generous man I have ever met – Anywhere, anytime. He is truly a brother by another mother. His family and my family have stayed close throughout the years.

We have shared several hunting trips in the USA, and are planning another this autumn. We have also been discussing another outing to Mongolia, if our advancing age permits us.

In 1986, Masoud's eldest son passed away from a heart attack while living in Salt Lake City, USA. To give an example how God's blessings have rained down upon us, when he called from Holland to England where I was then living, to tell us about Saeed's death, I was holding a ticket for me and my family to return to Salt Lake city on annual leave the next day. Three days later I was able to conduct Saeed's funeral in Salt Lake City, in the manner of his adopted religion, as well as his family's religious traditions. That opportunity to interact has been a blessing to us since the day I met Masoud Akhbari. The name Akhbari means the great one. I testify that this man lives up to his name; Masoud the Great One.

In 1977 I left Iran, to return to build a business in Logan, Utah. I used my Log Analysis skills to create a company to verify, and witness the acquisition of well logs, and then to analyze those logs.

In building this company, I had to have a computer system to use to process the tremendous volume of information acquired from the downhole acquisition. The only computers available in 1977 were beyond my means, so I undertook to build my own.

Micro-Processors were only just beginning to be available. They were still expensive, cranky, and memory speed and shortage were a major problem. I spent a year in research and development, and came up with a Log Analysis and digitizing system capable of doing all the processing needed for my business.

I had a two megahertz (2 million cycles per second) CPU, and 48 kilobytes of Ram Memory. This was coupled with a total of 600 Kilobytes of floppy disk storage. I had a dot matrix printer, and a monitor. The Digitizing system was state-of-the-art. It used two linear microphones at right angles, and with the known velocity of sound recognized a snapping sound from the digitizing pen to mark its' position.

This system was coupled with the software to put it all together and make it work. What a joke compared to systems of today. I carry more computing power and storage on my cell phone today, than was available in all the computers in North America at that time.

In 1978, Superior Oil Company was in need of a remote system to handle their oil exploration in Israel. They purchased my company from me, and as a consequence, hired me to go to Tel Aviv, Israel to conduct Geophysical, Petrophysical, and some Engineering operations for them. We moved the computers, hardware, and software to Tel Aviv, in late August of 1978.

The Camp David accord of 1978 put an end to oil exploration in the Gulf of Suez, and the Sinai, where Superior Oil Company was conducting operations, and so we began preparations to leave Israel.

In 1979, on November the 4th, the American embassy in Tehran was invaded, and hostages taken. This turmoil was felt all over the Middle East. At that time, Masoud was conducting his business in Iran, and preparing to move to Holland where his family had re-located. That period of relocation continued during the revolution of 1979, while Masoud was trapped in Iran. In early 1979, Masoud was able to leave Iran and move to be with his family in Holland and build his new business.

We took summer vacation that same year and visited Masoud and Willie with their two children in Holland.

From Left to Right standing; Heather T, Blaine, Judith T., Mandy A, Willie A. Seated; Teresa T.,Holly T., Fiona A., Masoud Tracy T. was taking the picture.

At that time, Masoud and I couldn't carry on our hunting together, so we just enjoyed our family. We were able to participate in a little "shooting" together by accident.

We were on a day trip to Scheveningen, which is a port city in Holland. They have an amusement park there, and we went into an arcade there where they had an air rifle and a target to shoot at. Masoud and I immediately gravitated to the air rifle, to reminisce about the air rifles we had in Iran and some of the memories associated with the discovery of this arcade. The proprietor soon became agitated with our taking up of his shop space, so Masoud kindly offered that I should try my hand to see if I could still shoot an air rifle.

I picked up the air rifle, which was firing .177 caliber steel air rifle shot, and fired at the target. Imagine my surprise when the target fired back a flash at me. I was quite surprised, and so was Masoud, and the proprietor!

He told us that there was a flash camera attached to the target. There was a sharp pin behind the bulls eye, and if this round, air rifle shot hit the sharp pin exactly straight on, the camera would take a picture of the shooter.

Masoud asked the man where the picture was, and he replied that in all the time he had the arcade no picture had ever been taken, so he had taken the film out of the polaroid camera, so as not to waste it. The flash was still connected up, but no film. He quickly sent his assistant out to get another box of film, and when he returned, the proprietor told me to hold the rifle back up, and took a staged picture. We had many a laugh at the camera with no film.

Camera with no film

We spent a week with the Akhbari family, and took day trips to Denmark and Friesland in the north of Holland. I seems that Masoud had a representative relationship with a company in Friesland that made the wonderful Dutch Clocks. Judith and I wanted to buy one, and so we travelled up to Masoud's company where we got a clock which we still have today.

In December 1979, our family relocated from Israel to Houston, Texas. We lived in the Woodlands, Texas, where I worked, and my children attended high school. The kids hated high school there so much, that it was an almost daily occurrence that they would beg me to go back abroad and get them out of the American public school system.

We only lived in the Woodlands for a little over a year, and then Conoco hired me to go the England in the spring of 1981. Finally we were back in Europe. It was a wonderful time for everyone. Now I could see my beloved friend and his family several times a year.

Every year we took a short vacation during the Easter holiday, and rode the ferry to the continent. The Keukenhof (Kitchen House) gardens near Amsterdam are beautiful in the spring of the year. Our yearly pilgrimage there helped to make the whole family appreciate the beauty of nature.

Keukenhof gardens with windmill in background

Hunting/Fishing in Scotland

While living in England, I discovered a hotel in Dumfriesshire, Scotland, which had a large holding of hunting properties, and a tremendous amount of river frontage which they had rented for fishing. Masoud and I made several trips to the Corsemalzie House Hotel in Newton Stewart, and enjoyed the salmon fishing and shooting there. In the autumn we would take a weekend and go there to shoot pheasants. On one occasion, two of my colleagues from Conoco joined us for a memorable pheasant shoot there.

Masoud and Pheasants at Corsemalzie House Hotel

Wes Rice, Masoud and Robert Moyer with Pheasants and Ducks at Corsemalzie House Hotel

Blaine sightseeing in Newton Stewart

Masoud in Dumfries

Hunting in Austria

Masoud purchased hunting rights with a group in Austria, near the villages of Uber Vaart, and Unter Vaart. We also made many trips there hunting, and enjoying each other's company. There was a "jagermeister" or hunting master there, by the name of Louis Mayer, who also served as "water gillie", or game warden. He took us on several hunts with Masoud, Blaine, Masoud's friends and colleagues from Holland, and our mutual friend from the USA, Robert Moyer.

Masoud and Louis Mayer; the Jagermeister examining a fine "Drilling" rifle/shotgun combination.

Robert Moyer, Masoud and Blaine at the hunting cabin in Austria

The Austrians really take their hunting seriously. While there in Ubervaart, we had an experience that was quite astounding to both of us. After Louis Mayer found that we were accomplished hunters and shooters, he took us on a hunting project; we were to shoot red deer for a goulash dinner at the church! Ubervaart is close to the Hungarian border, and during the autumn, they have a celebration of hunting. They make a big Hungarian Goulash dinner, and then invite all hunters, their families, and generally all interested people there to celebrate the hunting for the year.

The project is started on a Saturday morning, by assembling all the hunters for a field shoot, where they take pheasants, ducks, rabbits and quail. The hunt is a big one, with only shotguns, and small shot. They want to ensure there are no accidents.

They take an all-terrain vehicle and load it with pine boughs to hold the game, and go to the field. The shooters surround the small brushy areas in the fields, and beaters made up of the townspeople, who want to enjoy the goulash, beat the brush. Out fly pheasants and quail. Rabbits are also taken. All the animals are taken directly to the MOOG (all terrain vehicle), right after the field is cleared. In the MOOG, there is a 10 gallon bottle of peppermint schnapps. The schnapps is freely consumed by everyone whom has shot game.

The reason the game animals are taken so directly to the MOOG, is because the Germans and Austrians have a tradition of treating each animal with respect, which is almost reverence. At the first opportunity after the kill, a green sprig is placed in the animal's mouth. This is called the "letzebissen" or last bite. It reminds me of the "Halal", or ritual slaughter of clean animals in Muslim countries.

Bringing the animals for the "letzebissen". Masoud on the far right examining a pheasant.

Results of a good hunt!

Since Louis knew where there were red deer which needed to be thinned out, he took me out late on Saturday, and we collected two red deer cows, which didn't have calves with them. Red deer are very much like a North American Elk.

By the end of the day, the shooting is not as accurate as it began in the morning. If there is insufficient game collected by Saturday evening, then the store of meat must be supplemented by red deer, and roe deer.

Because of the schnapps consumed, the meat must be supplemented every year. When Louis found out that I didn't drink, he knew immediately who must collect the red deer meat for the dinner the next day.

On Sunday morning, all the "interested" people go to the Catholic Church for a ceremony, or Mass. This year a young hunter was to be honored in the ceremony. He had successfully passed his studies in order to eventually receive a hunting license. Under supervision, he had taken his first game; a rabbit. The priest then took a small willow branch, and with a ceremonial chant delivered three light strokes to the young man's upended buttocks, saying; one for the game, a second for the wild, and a third for Hubert (the patron saint of hunters).

After the ceremony, and Mass, the congregation was dismissed, and went to a local gathering hall for the afternoon's events.

The afternoon events included some more schnapps, and conversation, topped off by all the Hungarian goulash we could consume.

Robert Moyer, Louis Mayer, and Masoud examining the game animals being readied for the Goulash dinner

Blaine and Robert Moyer join in the Goulash Celebration

Schnapps, Coca-Cola and good friendly conversation started the afternoon

Our Austrian hunting colleagues examining a pheasant

Blaine and Masoud at the cabin in Austria

Robert Moyer, and Louis Mayer with the wild boar Robert shot in Austria

Masoud and Robert Moyer shooting clay pigeons at the local trap range.

The hospitality of the Austrian people is without question, the most wonderful thing about our trips to visit them there. Even though we had a comfortable cabin with a nice wood burning stove to cook our food, we were constantly being invited out. Dr. Kernbauer who is head of the hunting society for Untervaart invited us to dinner at her house on several occasions.

Masoud, Robert Moyer, Dr. Kernbauer, Louis Mayer, and Dr. Kernbauer's son

When we left Untervaart to drive back to Holland at the end of this trip, we thought we had experienced all the hospitality which we could bear. Little did we know.

When we got close to Vienna, which is one of my favorite cities in the world, we noticed that we were running low on fuel for Masoud's Mercedes car. We had been driving on secondary roads, and because everything closes early on Saturday evening, we hadn't found a fuel station. We didn't know that they were open 24/7 on the motorways, so we drove into Vienna to get fuel.

I have been to Vienna many times, but never driving my own car. I always flew in, or took the train in, and then used public transportation or taxi to get around. Vienna has a marvelous tram system, and if I lived in that city, I wouldn't even have a car.
We couldn't find a place to buy fuel. All the fuel stations are underground, and most were closed before midnight on Saturday evening.

We finally saw a young woman using a telephone booth, and we asked her where was the "benzene station". She replied back to me in machine gun rapid German, which I couldn't understand. When she saw that I didn't understand, she said to me in English; follow me, I know you won't find it by yourself.

She quickly drove to the benzene station under the central plaza in Vienna, and when she stopped to tell us how to use the pumps, Masoud grabbed the nozzle, and quickly filled her car as well as ours. What a way to say thanks. Masoud is so kind and gracious to everyone, everywhere.

I have thought many times about this experience. I know that everyone has their prejudices, both good and bad. It is interesting to me how our first strong experiences with both individuals and cultures form an impression that lasts. I guess the old saying is true; you only have one chance to make a first impression.

After filling up, we headed back for Holland. The motorways in Austria and Germany are something else. The motorway speeds are marked as "nicht sneller als" or not faster than 130 km/hr. This is an advisory speed limit only of 80 mph. When we arrived in Holland, Robert and Blaine flew back to England.

Unforeseen tragedy

In 1984, I was surprised to receive a phone call from Masoud's secretary. She told me that Masoud's eldest son, Saeed who was living in Utah, had suddenly and tragically passed away while eating dinner at the university restaurant.

Masoud had asked her to call and tell me that he was going to Salt Lake City, Utah, and handle affairs. I told her to "please call Mr. Akhbari at the Schipol airport, and tell him I would meet him the next day at the Salt Lake City Airport at 17:45". She told me that Masoud had anticipated that I would go to lend a hand, but he didn't want to trouble me. I told her that I was already holding tickets in my hand to take my family to Salt Lake City the next day.

It is uncanny how many times Masoud and I have been there for each other. This time was no exception. My own belief system tells me that this is more than just chance. Others can believe what they may like. Facts however are facts, and only reasons are up to doubt.

I arrived in Salt Lake City, and because I was serving as a Mormon church leader in England, we were afforded all the help and support by the local church officials as it was required. It was not that we were anything special, but we knew what to ask for. We were able to have a very nice funeral service, and bury Saeed's remains there in his newly adopted home country.

Every culture has its' own ceremonies and comforts centered around things like weddings, funerals, and births. These ceremonies ease and stabilize peoples' emotions at these tumultuous times.

Saeed had a lot of friends from the Mormon culture, because he had adopted that religion, but he also had a lot of Muslim friends because of the consequences of his birth and family ties. These two cultures have very differing ceremonial rites.

These ceremonial rites have some aspects which are very important, and other aspects which are just, normal. The important aspects should be looked after under any circumstances, while the "normal" ones can be passed over.

Saeed's funeral fit this category completely. Because I was very familiar with both sets of rites, I was able to conduct the funeral, and satisfy both sides of their requirements. I was even able to deliver the eulogy in Farsi, after 9 years of only an occasional brush with the language, when speaking with Masoud, or other Iranian friends.

We took these pictures of Saeed's final resting place about two years after his passing. We were amazed to find some "guardians" over his grave.

Saeed's grave marker, and his "guardians"

Missed opportunities Masoud has always been such a keen hunter and explorer. He never missed a trip. He is also an excellent businessman, so that his opportunities are great. In 1986, he asked me to accompany him on a hunting trip to Mongolia. The trip would require three weeks, and cost about $10,000 each.

I was fully engaged with a very demanding managerial job in England for Conoco, and my church duties as Bishop would make the time very demanding. I also had two children in university in the USA, and so I told Masoud I would have to pass.

Masoud in his great generosity realized that the financial burden I was facing was heavy, and he offered to pay for my trip. I think, in retrospect, that my pride was an over riding characteristic part of this decision. I am sure I could have found the time, and replacements to take over my responsibilities, but I didn't go. My father had a saying: "To test for indispensability, put your finger in a pail of water, then pull it out and see how big a hole you leave". "The truly indispensable will leave a hole, the rest of us will have to be satisfied with a ripple".

I really wished that I had taken the time to go on this trip. After returning, Masoud told me it wasn't much fun for him to go alone. I am ashamed that I wasn't there for my friend and brother on this occasion. I will endeavor not to let it happen again.

Masoud went for Marco Polo sheep, and Mongolian Ibex. The Marco Polo sheep was named by, you guessed it, Marco Polo. It's official name is "Ovis Ammon Polii", and is the largest of the sheep family. The Mongolian Ibex is also first in class for its' size.

Masoud travelled to Ulan Batur in Mongolia, then out into the Gobi Desert several hundred kilometers to hunt. He stayed in Yurts and the only common language between he and his guides were the few words of English which the guides had.

After two weeks of hunting, Masoud collected two marvelous trophies, to bring home to Holland.

Masoud with his Mongolian guide and Marco Polo Sheep

Masoud with the Mongolian Ibex

Hunting in Idaho

In 1989, my family and I returned to the united states, and moved onto a small acreage near Houston, Texas.

Judith's family still lived in Idaho. Masoud and I decided that we would like to try hunting there. Idaho and Utah both have all kinds of game and states allow out of state permits without much restriction. Part of the reason for this is the money. They are only able to charge a resident who is incidentally a *Voter* a nominal fee. The non resident however is fair game themselves. I guess this is the same thing that is done for special taxes on hotel rooms and rental cars. Visitors don't vote.

So we went to Idaho for deer hunting. Judith's family live in Idaho, only about 15 miles from the Utah border. The first year we went hunting there, we were late making a decision, and all the Idaho permits were already gone. Utah, however had plenty left, so we decided to go to Oakley, Idaho, and hunt back over the border in Utah.

The next year we were more timely, and we hunted right around Oakley on the Idaho side. There were plenty of deer, but we had the most enjoyment from the people.

We were staying in the cabin owned by Judith's family, and it was such a pleasant trip.

The Smith Cabin in the mountains above Oakley, Idaho

Hunting in the mountains in Idaho

Masoud and his youngest Daughter Fiona fixing breakfast

Fiona with the fruits of the hunt

This is a Big country! Masoud and Fiona enjoying the mountains

Masoud with his deer

Blaine, Dennis Smith, and Adam Smith – Yes that's Santa!

Hunting in Colorado

In 1991, Judith and I purchased a mountain property in Colorado. It was located above Alma, Colorado at a mean elevation of about 12,000 ft above sea level. The property was an old gold mining claim, which comprised almost 40 acres. It had a cabin of sorts, and was served by about 1 ½ miles of private road. The county road was not maintained in the winter, so it was a summer only place, or at best accessible by snowmobile.

We had two springs on the property. One was developed with a 55 gallon fiberglass reservoir, which was situated about 300 ft. in elevation above the cabin, giving almost 150 psi water pressure. We had plenty of water when things weren't frozen up.

The cabin was 20 ft square when we bought it, with a 10 foot by 20 foot loft. We quickly added a 10 foot extension onto the back which had indoor bathroom, and storage space.

We also added a walk around porch on two sides of the cabin giving 180 degree views of the surrounding area. What a beautiful place it was during about 2 ½ months of the year. The rest of the time it was all white.

One of our first hunts together in Colorado, was in 1992. Masoud came from Holland, with our good friend Robbert Kruisman. Robbert had accompanied us on one of the Idaho hunts.

Robbert and Masoud, because they were late deciding that they were able to come that year, had only taken deer permits. Wes Rice, my son-in-law Bill and I had successfully drawn in the lottery system for Elk tags.

Wes with his Cake

Wes, Robbert, Masoud and Bill in the snow

We hunted hard for a week, and Masoud collected a huge 3 point mule deer buck. We also took two Cow Elk, and a small Bull Elk. When we took the animals in the back of my pickup truck to the meat processing plant, we had the truck bed full of wild game.

Masoud with a grand 3 point buck

Masoud on the front steps with buck

On another trip, Masoud brought his son-in-law Cees. Cees is a competent hunter, and we took him along using my rifle. Cees had studied in Holland, and had European hunting permits, so he was well trained in gun safety and hunting procedures.

One morning early, we spotted the largest Mule Deer buck I have ever seen, very close to the road, just walking away from us. It was Cees turn, and I couldn't believe it that he didn't shoot. The buck was only about 75 yards away from us, and just slowly walking away from us. It finally made one jump into the trees, and wasn't seen again.

Cees told us that he didn't feel comfortable with the shot, because he was unsure that he could make a clean kill with the angle he had. We were so proud of his decision to resign the kill, over his principles. That's a true hunter!

Wes Rice was a Conoco colleague, and mutual friend who retired from service about the same time as I did. He fell ill with altitude sickness, and we had to put him on oxygen and then get him down out of the high elevation quickly. Elevation sickness is potentially fatal, so we had to move fast.

My grandson Devin also fell ill, and had to be moved quickly, but he was only a small child, so we didn't want to over-react.

In 1998, Judith and I took a quick trip to the cabin from our house in Texas, but we didn't want to spend a few days acclimatizing in the Denver area. We drove straight up to the property. That night Judith fell very ill, and by then we recognized all the signs, put her on oxygen, and moved her down to Denver within an hour.

I was fortunate, never to have suffered anything more than bad dreams, and shortness of breath, but we decided that after three close calls we had enough. In 1999 we sold the place.

We did have some memorable hunts there, and lots of good times together.

Cabin in the spring

View of the timberline from cabin

In 1995 we put a deck around the cabin to enjoy the views

Watch for Traffic! Is that a deer crossing cone marker?

Hunting in Texas
I have been spoiled most of my life by having lots of public land to hunt on. Only in England, Scotland, and Austria did we have to worry about finding a place to hunt.

Texas is another matter. There is very little public land there, and what little is available is very crowded come hunting time.

The answer is to rent or purchase hunting rights. Texas is also home to the largest high fence private hunting ranches in the world. These ranches which range from a square mile, to hundreds of square miles.

The famed Y-O ranch in south-central Texas, has been in operation since the 1880's and is over 40,000 acres. That's over 60 square miles. They have imported all kinds of exotic game and provide hunting opportunities on a pay as you go basis.

In 2007 Masoud and I decided to try one of these hunting ranches. I invited a mutual friend, and one of my sons-in-law, Jason to accompany us. Masoud flew into Houston, and we all drove to Abilene, Texas for a few days to try the hunting at the North Morgan Creek Ranch.

I have such mixed feelings about the 2007 hunt. The game was plentiful, and it was not like shooting fish in a barrel, but the guide was not pleasant. He was low class in his language and manners. We did not return there, and would not recommend it to any friend.

We did however make the best of the situation, and enjoyed ourselves for the hunting.

It was my first opportunity to take my Son-in-law, Jason hunting. It was certainly his first big game hunt as well. I gave him my father's rifle, an old military conversion enfield in 30-06 with a 4 power scope mounted.

My other friend used my Remington 700 .243 with a 20x scope. Both of those rifles performed superbly.

Masoud however had his Savage 7mm Remington Magnum. I bought this rifle for Masoud in 1989, and it resides in my gun safe. This has been a dependable gun, but on this occasion, the scope mounting bases had come loose. The rifle shot fine when he first tried it on targets, but with the ride to the shooting area, it had come adrift, and was causing the point of aim to wander several feet at a distance of 200 yards.

Even with the problem of the loose scope, Masoud was able to take a nice Audad Sheep.

I don't really like Texas style hunting. It is like most hunting in Europe, in that it is all done from blinds. Masoud and I much prefer to spot the animals, and then walk to where the game is. Sitting in a blind waiting for an animal to come by on a trail, or to a feeding place is not as demanding, or challenging.

I sat with Jason, for his first shot, to determine if he was subject to "buck fever" when he saw his first game. He was hunting wild hogs. It was evening, and the sun was just going down. There were plenty of small hogs going by on the trail, and to the feeding places, but Jason wanted a bigger hog. Finally just before it got too dark to shoot, a nice boar came walking by at about 150 yards. I asked Jason if he was ready, and then told him not to shoot at the hog, but a spot of mud on the hog's neck. With that advice he shot the hog perfectly. Down it went without any suffering at all. It is also a good thing because the meat is so unspoiled. The guide wanted to cut the hog's head off, and throw the rest away, because he said it wouldn't be good eating. That was one of the bad advice he gave us, which we didn't accept. This hog was cut and wrapped, it was perfect.

Jason with the Hog

Masoud shot the Audad from his stand, and what a surprise, it ran away. We could tell it had been hit, because of the blood trail, but we didn't yet realize that the scope mount had come loose. It didn't feel loose, and Masoud said that while he didn't miss often, maybe it was his fault. I should have known better. When we went tracking the Audad, and killed it, we felt much better.

The next day Masoud and I went for his hog. The package we had purchased allowed Masoud and I to take a large game animal, and one hog each. We sat in a blind and watched for hogs for some time. We sat through the morning and didn't see anything which looked worth shooting. At about the time we were ready to give up, the right hog presented itself. I was sitting watching it through the binoculars when Masoud fired.

To my great surprise, he shot under the animal! I knew then that there was something wrong with the rifle. Masoud was also incredulous that he had missed. When we took the scope off the gun, we found the mount had come loose from the gun, and was getting more and more loose with each shot or movement in the truck. I was very surprised, because I always use thread locker under hidden mount screws.

Masoud and his trophy Audad

I had purchased a package with a North American buffalo, or bison, plus a hog. This trip proved to be the trip of the malfunctioning guns!

On the first morning, I was sitting in a blind waiting for hogs. I had hand loaded some 220 grain bullets for my .300 Winchester Magnum specifically for the buffalo, and I wanted to try it on a hog to ensure proper performance.

As I sat there with the rifle loaded, but with the safety engaged, I looked through my riflescope at the various hogs as they came by. I was looking at a possibility, and thinking about taking a shot. I had the rifle pointed at the hog exactly where I would shoot. The hog was about 120 yards away, and I had decided I would shoot it in the head with that bullet, so as not to destroy any meat, but still make an instant kill.

The gun went off with a boom, and I realized I had been touching the trigger. The safety appeared to be engaged, but the rifle had still discharged. The good news is that I had followed the primary safety rule; "don't point at anything you don't want to kill". The hog was dead. Shot exactly in the head, and it didn't even kick.

That afternoon, the guide took me to shoot the buffalo. We chased buffalo all over that ranch. As soon as we would get within range, they would take off running. I didn't want to take a buffalo while it was running, because of what it would do to the meat.

I also didn't want to take the guide's advice about which buffalo I wanted to shoot. There were between 50 and 100 buffalo on that ranch, and I quite simply wanted a good eating one!

Finally I was satisfied with one of the animals, and even though it was 200 yards away, and looking right at me, I decided to take the shot. I sat down on the ground, and when I was aiming, the animal turned its' head. This was just the shot I wanted, and I took it right where the head and neck come together.

That heavy bullet killed the buffalo instantly. It went full length through the buffalo, and we found it right under the skin on its' backside. The bullet was expanded to about twice its' normal diameter, and did a wonderful job. I was very satisfied.

Masoud and Jason Standing, with Blaine kneeling with the buffalo

That .300 Winchester Magnum has always been a superb weapon. It is built on an Interarms Mauser action, with a 24 inch barrel. It has fired thousands of rounds and is still very accurate. It will shoot 5 shots into a 1 inch circle at 100 yards, using good handloaded ammo, as long as the shooter does his job.

Jason wanted to shoot my gun, and when we got back to the camp, we set up some targets. Jeff and Jason are both good sized men, and after firing three rounds of that heavy buffalo ammo, Jason thought he had a bruised shoulder. No doubt this gun kicks when loaded this heavy. It would easily take any of the African big 5 except for elephant.

Masoud and I have a history about hats. In Iran, after a particularly bad hunt for ducks, I was telling Masoud I was annoyed because all the ducks were flying on the other side of the water. He was embarrassed, but I was only joking with him. I never blamed him for our success, as he never blamed me. As we were joking I took his cloth hat, which was a cloth camouflage color, and threw it up in the air. Noel Saffiri, who was with us, quickly took his shotgun and shot the hat in the air. Masoud has never forgotten that experience. In the picture below, you can see him with my cowboy hat. It is just a private joke, that now is not so private anymore.

Relationship that has lasted 40 years topped by a hat!

I really liked the way we took care of moving the game back to the camp. Jeff McKnight who is one of my close friends was shooting wild boar, but he demonstrated the lift which the operator had fixed to the front of the pickup truck. With a wild animal the size of a buffalo, this was a labor saving device, which was a marvel to see.

This lift used the hardware of a truck mounted snowplow, but with a steel grating attached to carry the animals. A full sized buffalo can weigh 3000 pounds, and even a small one like I shot will weigh over 1000 pounds.

Buffalo meat, in common with almost all wild game meat has almost no bad or ldl cholesterol. The only domesticated meat animals with that characteristic is the meat from the longhorn cattle from Texas.

We filled several freezers with meat those few days. Hogs, Bufalo, and Audad. Since the Audad had run after being shot, we used it with the hogs and some domestic pork to make sausage. What a wonderful sausage that was.

Jeff demonstrating the game carrier lift

Loading the Buffalo

After we returned to Richmond, from our hunt, we were able to get in touch with Gil Jurenka, whom is featured in an earlier chapter in this book. Gil and his wife came to meet us and made a nice reunion with Masoud. We have included a description of Gil Jurenka's adventures with Masoud in Iran. You can find some pictures from Masoud's library in Chapter 9.

THE JURENKA FAMILY IN IRAN

My name is Gilbert (Gil) Jurenka. I was a Senior Reservoir Engineer with Mobil Oil Company in June, 1965 when I accepted a loan assignment from Mobil to the Iranian Oil Exploration and Production Company (IOE&PC). A month later, I and my wife Shirley, and three young children arrived in Abadan, Iran at 4:00 AM on a flight from Beirut, Lebanon. Little did we know how interesting and varied would be our life over the next two years.

Later in the day of our arrival, we were placed on a flight to Masjid-i-Suleiman (MIS) where we were housed in a Company facility for about a week while our permanent home in the ex-patriot housing development called Naftak was readied for our use. Over the course of the next three weeks, we were assisted in acquiring a houseboy-cook, a gardener, an automobile and various other needs to complete our new home. We received an indoctrination regarding the local customs, availability of Company facilities, the city's bazaar, the local Christian church, recreational opportunities, schools for the children, etc., etc. By the end of those three weeks, I was established in my new job, Shirley had fairly well developed her daily routine and the children were readied for school.

It was immediately obvious to me that reservoir engineering in Iran was going to involve a very complex undertaking .... A project to finalize and test a computer model which would: 1) identify operational procedures needed to maximize ultimate oil recovery from the tight, but highly fractured Asmari limestone reservoirs, and 2) aid in forecasting future production for use in developing capital budgets and income projections. Much of the computer programming had already been done, but it was necessary to fine-tune the input data specific to each of the several oil fields that constituted IOE&PC's Iranian concession. It was a challenging job.

Work hours were from 6:00 AM to 1:00 PM during the summer months with a one-half hour break at 10:00 AM. During the spring, fall and winter, work hours were from 7:00 AM to 4:00 PM with a one-hour lunch break. The work week was Monday through Saturday but Saturday's work ended at noon during the spring, fall and winter. These work schedules obviously left plenty of time for recreation, which for me meant a lot of golf.

Golfing was a major sport for both ex-patriots and a good number of Iranians. The ex-patriot golfers consisted of English, French, Scots, Irish, Dutch, Australian and American citizens. Many of us played almost every day, even during the summer when temperatures often reached 115 degrees. There were a number of competitions within the local club as well as with other IOE&PC communities in the southern part of Iran and also Tehran in the north. Largely due to playing a lot of golf, I was fortunate to win the Khuzestan Open Golf Tournament held in MIS in 1967.

All golfers ha their own personal caddies who were primarily young men and boys from the local population. Some years after our return to the United States, I was thrilled to get a phone call from my caddie, Ali, who was attending college in Lafayette, Louisiana, probably under the sponsorship of another American golfer who worked in Iran.

Shirley, like many of the ex-patriot wives, had only one serious problem to deal with … tedium. Our three children were young and required a good deal of supervision, oversight, shopping for food and clothing and school involvement. The only breaks from this daily routine came from a round of golf every once in a while, visits to other wives, evening movies at the company theater, and frequent parties hosted by friends and/or the company (for visiting dignitaries). There were very few trips outside the area of MIS. She did experience on life altering event … the birth of our third son (fourth child). Shirley had no trouble with the delivery, but had trouble convincing the nurses that she did not have an Iranian husband. I was so tanned from playing a lot of golf in the sun, that they just knew I had to be a native.

For me, an avid hunter, many of the unforgettable memories of the time spent in Iran originated from hunting trips. Two factors that made that possible were first, my early introduction to and friendship I developed with English-speaking Masoud Akhbari who was head of Drafting and Reproduction for IOE&PC. Masoud too was an avid hunter and he knew how to communicate with and arrange hunting trips with the local hunters and Bakhtiari tribal people who populated the rural areas around MIS during the winter hunting months. While working at my desk, I would often sense someone standing at the door and looked up to see a Bakhtiari in his full tribal dress. He would simply say "Mr. Jurenka…chucker". I knew that this meant wild game and I would get up and take him to see Masoud who would interpret his report about sighting some game that he wanted me to come hunt. Within a day, Masoud would come to me with access to a land rover and a plan to make the hunt.

The second factor providing my ability to hunt in Iran was the availability of a shotgun and rifle, together with the appropriate permits, that were loaned to me by Jug Bell, another American IOE&PC employee from Victoria, Texas. Jug had imported guns and obtained the permits before such activity was banned by the Iranian government (by coincidence, Jug had officiated football games in Vanderbilt, Texas at the time Shirley was drum majorette for the high school band). Obviously, somehow the word got around that I had access to the guns and I believe no one loves to hunt more than the Bakhtiari.

The hunts that will forever be imprinted in my mind are:

- An Ibex hunt (unsuccessful) involving a difficult 5 hour climb to the top of an exposed Asmari anticline, and an equally difficult decent.

- A duck hunt (also unsuccessful) in the marshes near the Caspian Sea which involved a 3 hour ride by boat, rowed by a local Iranian who, in spite of the cold, had nothing but thin socks and loafers on his feet.

- A chucker hunt (highly successful) near Gachsaran that included the Governor General of Khuzestan Province, the General manager of the IOE&PC, two other managers, and me (remember, I had the gun and the permit). This hunt involved a helicopter ride into the foothills of the Zagros Mountains and an encampment well-guarded by Iranian Gendarmes because of the history of violence from the local tribes.

- A Mouflon sheep hunt with Masoud in the Zagros Mountains near Isfahan. This hunt involved a night time climb by horseback. We encountered a herd of sheep with several rams about 4:00 PM the second day and I took a nice one. One of the tribesmen who were guiding us gutted the sheep and another one put the sheep carcass on his back and we proceeded on a fast trot back to the village in order to get there before dark. The Bakhtiari were incredibly strong and hardy people.

- A chucker hunt near MIS with Masoud where, while hunting on my own, I encountered a Bakhtiari tribesman who was also hunting chucker (had a wired up Browning automatic). We hunted together for an hour or so after which he pointed to their encampment about a mile away and indicated I should follow him. Upon arrival there, the women and children all gathered blankets, rugs and pillows for me and brought me a cup of bitter tea with sugar clumps. It was the only tea I have ever drank in my life. I remember the entire group pointing to my clothing and shoes and laughing. I was a star. Masoud finally found me and ruined their entertainment.

These hunting adventures occupy a large part of my memory of our time in Iran, but there are a great many other "stories" to tell about the people, places visited, the environment, and the work. I really would not know where to begin … or end … telling them.

We truly enjoyed all the Iranian people with whom we worked and played while in their country. Some of the local teenagers infrequently displayed a bit of resentment to our presence, but we were met with nothing but respect from shop owners in the bazaar, tribes people, and fellow Iranian employees. We particularly enjoyed the times we shared with many of the ex-patriot community, and our children still frequently recall their school experiences, the fun they had with Esau, our houseboy/cook, and Ali, our yardman and gardener. They were good people and we will never forget them.

We did not sign up for a second tour of duty in Iran, not because we were unhappy there, but because my career aspirations were to seek a management position with Mobil, a goal I thought might best be achieved via a more visual position in domestic operations. Masoud, his wife and several of his local Iranian friends, with whom we had enjoyed hunting expeditions, hosted a wild pig roast and party before our departure. It was a sad farewell knowing it was likely we would never again see most of the people who were there to see us off.

In the early summer of 2007, Shirley and I received a wonderful surprise in the form of a telephone call from Masoud asking if it would be possible to meet him in Houston where he was planning to visit another American friend, Blaine Taylor. We were so happy for this opportunity that we would have cancelled any existing plans to meet such a dear friend. Fortunately, the timing for such a visit fit very well with our annual trip to our fishing camp in Louisiana. What a happy day, some 40 years after last seeing Masoud, that we would again be able to see and visit with him.

The meeting took place at a restaurant just off Highway 59 on the southwest side of Houston. There are no words to describe how elated we were to extend our arms for a hug from a man that was such an important part of our Iranian adventure. Needless to say, it took a few hours for each of us to attempt an update on our respective lives covering nearly a half century. Of course, it was also necessary for us to recall a great many memories we shared, particularly concerning our hunting expeditions. Shirley and I both ended up with wet eyes after saying what likely would be, this time, our final goodbye.

Gil and Masoud in Houston, Texas

Masoud and the Jurenkas'

Other Texas Hunting

A few years ago, we went night hunting hogs with my grandson Taylor and his younger brother Devin. Taylor was using my rifle, and I was spotting. The boys have been taught well about marksmanship, and gun safety, but those first encounters with live game are different than a trip to the shooting range.

Taylor shot a hog from a stand at a range of 100 yards, and hit it precisely in the head. Devin was a little more excited however, and missed his opportunity.

A year later, we took Devin on a hog hunt in Central Texas, after many sessions of gun safety, training, and target shooting. We were doing my favorite kind of hunting, where we were walking up the hogs. Devin shot a nice hog, and then I took one. I asked Devin if he would like another, and he said he would. (What a surprise)!

I had reminded Devin on several occasions to be careful of his back-stop, because in the event of a miss, or even a complete penetration, you needed to know where that bullet was going. He fired, and killed three pigs with the one bullet. Devin now knows why you must be careful of the line of fire!

We went home that day with 5 hogs in the cooler. My daughter had to buy a new freezer to hold it all.

My Children and grand-children, and even some great grandchildren are being raised with the old western traditions. All four of my children are quite competent with weapons, as are their children. Hunting is an excuse for getting together.

Judith however doesn't much like camping. She never has. When asked about camping she will tell you that: "it took man two million years to get out of the caves, and I for one am not going back". Everyone else likes to camp in reasonable weather.

Like many people, when you were young and struggling financially, family vacations had to be camping vacations.

Masoud and Ruud Paauwe in his "Trophy" Room at home in Holland

Future Plans

In December 2012 Masoud will travel to Utah, USA, with Fiona his daughter to accompany Blaine and his son-in-law on a deer hunt.

We have a very good friend of the family who has arranged for a deer hunt on his property. Because of the drought conditions in the west this year, there are too many deer for the feed which is available to take them through the winter and early spring. The state has granted land-owner permits which allow us as colleagues to go hunting.

In 2013, Masoud is planning another hunt to Mongolia. I am not going to miss this one. This hunt is planned as a celebratory hunt for the culmination of a large business deal Masoud is culminating with his son who lives in Iran.

The Akhbari family, with Masoud, Willie, Daughters Mandy and Fiona, and the Grandchildren

Forty years later, two old men, who are brothers of different families walk a snowy street half way round the world, wondering what's around the next corner.

The Saga Continues.